WASTE AND POLLUTION

WASTE AND POLLUTION

The Problem for Britain

Kenneth Mellanby

HarperCollins*Publishers*

HarperCollins*Publishers*
London · Glasgow · Sydney · Auckland
Toronto · Johannesburg

First published 1992
© Kenneth Mellanby 1992

ISBN 0 00 219182 2

Printed by Butler & Tanner, Frome, Somerset, UK

Contents

Preface	7
The Problem of Waste	9
Finding Sites for Waste Disposal	15
Municipal Waste Sites	23
Mining Wastes	43
Coal mining	44
Iron ore	47
Sand and gravel	49
Clay for brickmaking	50
China clay	52
Quarries	54
Hazardous Wastes	55
Air Pollution	69
Water Pollution	87
Pesticides	109
Herbicides	112
Fungicides	114
Insecticides	115
Waste Heat	121
The Greenhouse Effect	129
The Ozone Layer and Ultra-violet Radiation	141
Radioactive Wastes	148
The Future	161
Appendices	165
Bibliography	179
Index	187

Preface

In the crowded British Isles every part of the countryside, and of the urban area, is affected by man's activities. This volume is about the way our various waste products, and the methods used to deal with them, contribute to this effect. As an ecologist who has been for many years concerned with the conservation of wildlife, I have described the impact of waste products on our native flora and fauna, as well as the ways in which man and his environment is affected. It has been shown that various species of wildlife are particularly susceptible to damage from toxic substances, and they may give us an early warning of possible dangers to man, in the same way that canaries were taken down coal mines to warn the miners of air pollution.

In many cases serious damage to the environment, and to individual species, is obviously occurring. Landfill sites, used for the disposal of both domestic and industrial waste, may cover areas of agricultural or wildlife importance. Sites previously damaged by developments such as quarrying or gravel extraction, if left undisturbed, often develop considerable conservation interest, being invaded by rare plants and providing habitats for birds. So, even where the original exploitation was only permitted provided the site was restored to its previous use (generally agricultural), and where it was stipulated that domestic refuse would be used to fill the cavity caused by the mineral extraction, conservationists have often opposed such restoration, and have preferred to retain the unimproved landscape.

As a rule, wastes are better dealt with today than at any previous time, and they cause less pollution and environmental damage than ever before. This is to our benefit, but it may not be so appreciated by those concerned with wildlife. Thus rubbish dumps have now been replaced by sanitary landfill sites, which deal hygienically and safely with municipal and other wastes. This improved management means that they no longer serve as important food sources for gulls and other birds, and also that they no longer support such a rich fauna as did the old badly managed refuse tips. Sewage treatment is pure and safe, the sites almost clinical in appearance. But we have lost the 'sewage farms' which were famous for their birdlife, including rare and interesting migratory waders. Toxic chemicals are now isolated and often destroyed by treatments which yield no dangerous residues. This is generally to be welcomed, but never again will we produce the wonderful

8 WASTE AND POLLUTION

sites with their outstanding populations of orchids which have developed where some of the most dangerous chemicals were dumped, with no concern for the environment, in Lancashire towards the end of the 19th century. The soil was sterilized by toxic waste, and conditions which prevented other more rampant plants from growing were created – a classic case of serendipity.

My main purpose in writing this book is to try to give a balanced and accurate picture of what is really happening to our wastes, and how they are affecting our countryside. Unfortunately, it is not always easy for the public to obtain such a picture. Reports of doom and destruction are more exciting than those which give a less pessimistic picture. The media give too much space to those whom the French politician M. Jean Valroff, in his report *Pollution atmosphérique et pluies acides,* calls *'charlatans de l' écologie'*. Thus we often get greatly exaggerated accounts of damage, with misleading attempts to explain its causes.

We also suffer from inaccurate forecasts of the future, sometimes made by scientists who should know better, often by so-called 'ecologists' who deserve M. Valroff's description. Among scientists, Professor Paul Ehrlich of Stanford University, who is a distinguished authority on lepidoptera and their populations, said in 1969 that before 1980 the world's oceans would be dead, and that the Chinese and Japanese would suffer hardship from the disappearance of fish and other sea food from their diet. Having made an extensive tour to the Far East in 1988, I am glad to report that no such catastrophe has occurred, and that I enjoyed many excellent meals of sea food. Then in *The Blueprint for Survival,* published in 1972 in *The Ecologist* magazine, we were told that global supplies of silver, gold, mercury, lead and zinc would be nearing exhaustion by 1987. In fact, there have been gluts rather than shortages, and prices have fallen so much that research on how to extract metals from low-grade ores has been largely stopped. The same *Blueprint* also stated that Britain's farms would suffer substantial decreases in productivity, causing food shortages, when in fact in the ten years after 1972 there was the greatest increase in crop yields ever seen in all our history. I hope that I have not made any similar prognostications.

Without minimizing the seriousness of the situation, I wish to point out the difference between the real and imaginary damage and to show, where possible, how matters could be improved. We need to learn to distinguish between genuine damaging pollution and cases where toxic substances are present in such small quantities that they are harmless, or, in some cases, actually beneficial. Resources are always limited, and we must concentrate our efforts on removing the causes of real damage, rather than frittering away these resources on unnecessary efforts to eliminate harmless contaminants of our environment.

Chapter 1

THE PROBLEM OF WASTE

Waste, refuse, rubbish, garbage and trash are all words used to describe something unpleasant – and something produced by man alone. The problem of waste, and waste disposal, is therefore entirely man-made. Before *Homo sapiens* appeared on earth, there were no real waste materials, although even our most distant ancestors soon saw to it that a waste problem was created.

The body constituents of most plants and animals are efficiently recycled after they die. The dead organism is rapidly broken down by other animals, by fungi and by bacteria so that its essential chemicals are liberated into the soil. There they serve as nutrients for growing plants, which in turn are consumed by herbivorous animals. Most of the nutrients are then quickly returned to the soil in the herbivore's excrement. The herbivores themselves also serve as food for carnivores, which similarly recycle nutrients in their excrement. Finally, the bodies of the carnivores decay after their death, and their substances are incorporated into the soil for further use by the plants.

In the wild, it is unusual for either dung or corpses to accumulate in objectionable amounts, because the organisms concerned with recycling are generally ready and waiting to perform their essential duties. The only exception is when alien introductions produce dung to which the native consumers are unadapted. Thus in Australia, the insects which dealt efficiently with kangaroo faeces could not break down cowpats, which then produced swarms of objectionable flies. Today, dung beetles from the countries which were the homes of the cattle have been introduced into Australia to reduce this nuisance. On the other hand, we have often made the recycling of man-made wastes a slow, difficult and unpleasant process – the situation being greatly complicated by the sophistications of civilization.

Decay and recycling is the rule for most dead animal and plant materials, but under certain circumstances other processes take over. Our so-called fossil fuels – coal, oil and natural gas – were produced many millions of years ago when the dead vegetation did not decompose. We can see the beginning of the processes which produced these energy-rich fuels in a modern peat bog. Here the acid conditions inhibit decay, and the plant material is preserved in the form of peat which, with compression and time, could produce coal. It is true that other substances, including animal matter, may also be preserved in peat bogs. Archaeologists have found wooden artefacts, textiles and even bodies (Tollund man was dug up in Denmark, and a similar corpse from a Cheshire bog in a remarkable state of preservation). However, these are special cases – breakdown and decay is the general rule.

The distinguished archaeologist Jacquetta Hawkes writes 'Waste disposal by human beings may be said to have begun when hunters of the earliest Stone Age tossed their gnawn bones over their shoulders.' She points out that Palaeolothic hunters 'were not cave-proud. They allowed bones and other animal debris, wood ash and waste flakes of flint knapping to accumulate under them until the cave might be reduced to half its height.' So their methods of disposal were not very efficient, and most ancient settlements were eventually buried under their own debris.

Excrement, dung and urine, have always produced hygienic and aesthetic problems for man and, in fact, for other animals. Birds are said not to 'foul their own nests' and may take precautions not to do so. Thus the young songbirds in the nest may produce their excrement in sacs which are tidily removed from the nests by the parents and dropped some distance away. Many mammals, including rabbits, have areas away from their dens which serve as 'latrines', where the ground first shows signs of great fertility and then, with continued use, of pollution from over-fertilization, which kills the vegetation.

Nomadic people were, and often still are, unhygienic, moving their camp sites when they become too unpleasant. However, the ancient Israelites were aware of the problem. In Deuteronomy, Chapter 23, verses 12 and 13 we read: 'Thou shalt have a place without the camp, whither thou shalt go forth abroad; And thou shalt have a paddle upon thy weapon; and it shall be, when thou wilt ease thyself abroad, thou shalt dig therewith, and shalt turn back and cover that which cometh from thee.' That is the Authorised Version. In this case I prefer the James Moffatt translation, first published in Scotland in 1924, which puts it rather more concisely: 'Also, you must have a latrine outside the camp and go outside to it; you must carry a spade among your weapons, and when you relieve yourself outside, you must dig a hole with it, to cover up your filth.'

Jacquetta Hawkes comments on London as follows: 'The city of London may be regarded as a modest form of tell (i.e. an artificial mound of accumulated remains of ancient settlements) with the streets of the

Roman town lying some 20 feet below our own. It provides a convincing proof of the greater civic efficiency of the Romans over Londoners of the Middle Ages. At both periods pits were dug and filled with waste, but while in Roman times these held ash and other dry materials, those of a thousand years or so later were used for organic domestic refuse that must have added to the notorious stink of medieval London.' Similarly, reports of excavations of Anglo-Saxon York show that much of the rubbish often remained within the house, where the floors were covered with material which had 'many resemblances to that of herbivorous dung', and which supported a similar fauna of scavenging insects.

The wastes produced by our unhygienic ancestors, and the materials found in their 'kitchen middens' have enabled archaeologists to produce a picture of their life-style. In the same way modern, investigative journalists, often in defiance of law, examine the contents of the dustbins of the famous or the notorious, and by counting champagne bottles or caviar jars, produce articles describing the conspicuous eating habits of their victims.

Until a hundred years ago, waste disposal and sewage treatment in London and other cities was very ineffective, although many of the great houses and the monasteries had reasonably salubrious privies. In the cities most refuse was allowed to rot in the streets, where it encouraged vermin of various types, including rats, mice and crows. Until the end of the 18th century, the Red Kite (*Milvus milvus*) was common in London, scavenging on the refuse. The presence of this species, now a rare denizen of the more remote parts of Wales, was resented by many of the inhabitants. They were probably wrong in their attitude, for kites got rid of stinking and decaying food, and probably discouraged rats, which were potentially more dangerous carriers of disease. Kites became scarce in the 19th century, when they ran short of food in the cities, and were harried by gamekeepers in the country. In 1900 only 12 pairs remained in Wales; the bird was extinct throughout the rest of Britain. The Red Kite is now strictly preserved, and its population has increased to over 40 breeding pairs. One reason for this increase may be that they have returned to the refuse tips of Wales (just as their relatives do in Africa and Australia), and this extra food supply may be as useful to them as it is to the hordes of gulls which also frequent the refuse disposal sites.

When humans came to live in towns and cities, there was usually some attempt to reduce pollution of the whole environment from excrement, although this was not always successful. Contemporary reports suggest that even such grand places as the Palace of Versailles stank rather badly. Attempts were made to alleviate the smell by the use of pomanders, and by spreading sweet herbs on the floor. Royalty and the aristocracy were constantly travelling around the country. These peregrinatory habits were partly to escape from the ordure and other noxious waste materials which the court always generated.

Until a couple of centuries ago, when there was no main drainage, chamber pots in Edinburgh were emptied from upper windows into the street; hence the warning cry of 'Gardy loo' (Scotch-French for 'beware of the water') so that passers-by could try to avoid being soaked with the contents. One of the earliest types of public lavatory, featured in a series of prints of London's street life, was a bucket carried by a woman, into which passers-by who were caught short could defecate covered by the woman's cloak (with only their head appearing rather like a young 'joey' in a kangaroo's pouch).

However, in the towns at any rate, much of the excrement was voided into some sort of latrine. James Greig has described the contents of a 15th century barrel latrine in Worcester, where the concentrated faecal matter was well preserved. Greig found traces of 20 different plants, mainly fruits and vegetables, together with a rich fauna of human intestinal parasites. Night soil from earth closets was known to be a rich fertilizer, and was generally collected (together with material left about the streets) and used in market gardens, producing excellent, nutritious if unhygienic and parasite-infested vegetables. The market gardens of Hertfordshire, for instance, were famous for their fertility, which was based on the use of London's night soil. It is likely that all members of the population, from the richest to the paupers, carried a selection of roundworms and tapeworms as well as many bacterial and protozooal parasites.

Actually, human dung is still employed as manure in many countries. Its use in China is particularly well developed, and here the material, perhaps because it is often partially sterilized by the heat when it is composted, appears to avoid some of the infestation from intestinal parasites common in other countries. Until recently, less effort was made to prevent pollution from urine than from faeces, and in Britain we still find notices in quiet passages in our cities urging people to 'Commit no Nuisance'. Where these notices are ignored, the plants, in particular algae and bryophytes, which grow on the walls, are likely to be affected. At first they will benefit from the rich source of plant nutrients, but will soon suffer from a surfeit of riches.

The introduction of the water closet provided the opportunity for improving the situation. This convenience is generally believed to have been invented by Sir John Harington, godson of Queen Elizabeth, in the 16th century. It is said that the Queen installed a model in Windsor Castle, and that she kept a copy of Sir John's masterpiece *The Metamorphosis of Ajax* (i.e. a jakes, that is a privy), which describes his invention, as reading matter in the appropriate room. However, like so many British inventions, this one was not widely exploited at home, and WCs do not seem to have been at all common until the appropriately named Mr Crapper went into business in London in the early 19th century. Flushing water closets were then widely introduced, with a great improvement in urban conditions; but as the contents of the WCs were discharged untreated into the Thames, the effect on the tidal stretches of the river was horrendous. At times Parliament had to suspend sitting because of the stench. During the summer the windows were covered with cloths soaked in disinfectant. The river became 'dead' and all fish and most other forms of aquatic life were eliminated. 1856 became known as 'the year of the great stink'.

The situation improved from the end of the 19th century onward. The capital's raw sewage was diverted from the London stretch of the Thames, and discharged further down the river (where, of course, damage continued to occur, but since the area was much less densely populated, fewer people, and no Members of Parliament, complained). Then various methods of sewage treatment were installed, not only in London but in all towns further up the river. By the 1960s fish were found to have returned even to what had been the most polluted reaches. The importance of 'sewage farms' and other forms of treatment are discussed later in this book.

Today we have a major industry disposing of our wastes as is described later in this book. Domestic refuse is regularly collected from practically every dwelling in Britain and in other developed countries. This has considerably reduced the food supply of urban vermin, although foxes may still pillage dustbins. In Britain landfill is much the most important form of refuse disposal. This is, in effect, a development of the refuse pits of Roman and medieval London. These waste disposal sites may develop considerable ecological interest, as was shown by Arnold Darlington in 1969 in his pioneer work *The Ecology of Refuse Tips*. The management of such sites obviously has a considerable effect on the animals and plants found on them.

To most of us, the 20,000,000 tonnes of domestic waste – sometimes called 'municipal waste' – constitute the main problem in Britain today. This is the material we put out of our houses every week in dustbins or plastic sacks, the collection of which adds substantially to our rates. It is what so disfigures many of our streets, where leaking sacks may be ripped open by dogs in search of bones, and where cardboard cartons add to the litter. However, as is indicated in the eleventh report of the Royal Commission on Environmental Pollution (Cmd 9675, 1985) it constitutes only about four per cent of the total waste of some 482,000,000 tonnes generated in Britain. Agriculture makes a contribution of 200,000,000 tonnes, mineral working of 135,000,000 tonnes and industry of 100,000,000 tonnes.

All these different types of waste produce different ecological problems. Recently there has been much concern about so-called 'toxic wastes', although many substances so classified are not particularly poisonous, and those that are may be present in very small quantities. Perhaps the public is most worried by the problem of the safe disposal of radioactive waste from nuclear power stations. The problems of disposing of these different wastes are discussed in later chapters of this book. Surprising results are sometimes obtained. Those wastes which might be expected to have the greatest environmental effects, such as radioactive materials, often do little damage, and the sites containing the most toxic substances may develop the most interesting flora.

In addition to the solid and liquid wastes produced directly by humans and by industry, we have many gaseous wastes given out mainly as a by-product of the liberation of energy from fossil fuels. When these gases are in high enough concentrations, they can do serious damage to living organisms and to whole ecosystems. As well as these direct effects, they may produce important results in areas far distant from the site of emission, as the gases may suffer chemical changes as they pass through the atmosphere. This results in the phenomenon commonly called 'acid rain'. As will be seen in Chapters 5 and 6, this is a much more complicated subject than is generally realized. Furthermore, these waste gases, and other emissions from industrial processes and domestic apparatus, may have global effects, possibly affecting climate and the amount of ultraviolet radiation reaching the earth's surface.

Waste heat, particularly from electric generating stations, affects many rivers, lakes and even parts of the sea, when water is abstracted and used for cooling purposes. This heat may have important consequences for fresh water and for marine flora and fauna.

There is public concern about the potential danger from radioactive waste material from nuclear power stations, and also from possible 'leaks' of radiation from the power stations themselves. The ecological effects of all such radiation have often been considerably exaggerated, but nevertheless this is clearly a subject which merits our attention.

The purpose of this book is to describe the effects of all important types of waste upon the countryside and the environment in which we live. I have discussed the effects on man, and on the world as a whole, and in some cases I have suggested ways in which both local and global dangers may be avoided.

Chapter 2

FINDING SITES FOR WASTE DISPOSAL

Civilized man produces a great deal of waste. As shown in the last chapter, domestic waste in Britain amounts, in a year, to some 20 million tonnes, with an untreated volume of 45 million cubic metres. County councils, who are responsible for waste disposal, must find suitable sites for the safe and hygienic disposal of all this material. This is by no means a simple task.

Domestic waste contains the following main ingredients:

Vegetable and putrescible matter	4 million tonnes
Paper and cardboard	6 million tonnes
Metal, mainly iron ('tin cans')	2 million tonnes
Glass ('cullet')	2.5 million tonnes
Plastic	1 million tonnes

The proportions of these ingredients have changed greatly in the last 50 years. Whereas coal ash was the major constituent at the beginning of this period, today there is almost none in the waste from cities like London which are mainly in smokeless zones, so that no coal is burned. Paper and cardboard have increased annually, as newspapers become more bulky, and more goods are packed in non-returnable cardboard boxes. This large amount of what is mainly cellulose is the cause of what is probably the most serious long-term problem related to waste disposal, namely the production of explosive methane gas, something which continues for many years after the refuse is deposited. Yet although it presents a problem, this methane can sometimes be collected, and then it becomes a valuable resource. Plastic is a recent and rapidly increasing ingredient, which by its bulk and indestructibility appears to represent a much greater fraction than these figures suggest.

The eleventh report of the Royal Commission on Environmental Pollution (1985) showed that, contrary to expectation, the total weight of household waste produced by the average household has not increased in recent years. It actually fell by about one-third between 1936 and

1974; this may largely have been the result of the disappearance of heavy residues from burning coal. Since 1974 the weight has remained nearly constant, although the bulk may have increased because of the larger quantity of light and bulky plastic containers and packing materials.

In Britain today, over 90 per cent of domestic wastes, and a substantial amount of industrial waste, goes into what the public call 'rubbish dumps', and which are more correctly called 'sanitary landfill sites'. The term used by the waste disposal industry is not just a case of an unnecessary euphemism (like the substitution of 'garbage operative' for 'dustman'); it describes the process of disposal more accurately, for the waste is carefully treated so as to reduce its nuisance value, whereas no such precautions are taken with a dump. There are indeed still some true rubbish dumps, where wastes are fly-tipped by irresponsible 'cowboys' and by uncaring citizens. The fly-tipped material accounts only for a tiny fraction of all domestic refuse, but it adds seriously to Britain's litter problem. Most local authorities, as well as collecting domestic refuse, operate some system, generally with a modest charge, for taking away large objects such as derelict refrigerators or broken furniture. They also have collecting sites to which people can bring their own discarded materials. Unfortunately, many seem to prefer to dump their junk on roadsides and are more than likely to choose sites of scenic or ecological value.

I have observed this process at Hares Down and Knowlestone Moor SSSI (Site of Special Scientific Interest) in North Devon. Here there was a prolonged Public Inquiry in 1983 and 1984 into the proposal that the North Devon Link Road, improving access from the M5 motorway to Barnstaple, should cross this important ecological site. When I first visited Hares Down I found a small area covered with refuse, including discarded household machinery and garden clippings. Over the next year this accurately named dump grew considerably. The local conservationists who were so vocal in their opposition to the road had done nothing to remove this eyesore, which they must have seen every time they visited the moor. This type of fly-tipping, incidentally, contains little in the way of food residues, so at least it does not encourage rats or birds which frequent refuse disposal sites.

Wildlife as a whole has little aesthetic sense, and some birds, even in a rural area with many alternative niches, use abandoned refrigerators or broken-down cookers for nesting sites. Thus at Hamilton in Victoria, Australia, Gunn's Bandicoot (*Parameles gunni*) survives and flourishes in old car bodies on one tip, and there was an outcry from conservationists when it was proposed to tidy up the site and remove the cars. In

California and other coastal areas, derelict automobiles dumped offshore provide refuges for lobsters and other forms of marine life.

Refuse may supply food to wildlife when it is put out by householders and before it is collected by the local authorities. There are many reports of urban foxes and feral cats raiding dustbins. Plastic sacks containing food waste may be punctured and the contents spread over the street, to be consumed by rats, sparrows and pigeons. Houseflies and bluebottles lay their eggs on unprotected refuse; the eggs usually hatch on the refuse tips, and the flies may emerge there later on. However, most waste food is duly collected, and transported to disposal sites, where it gives rise to further problems.

Sites have to be found where all this material may be deposited and properly managed. Every year 18 million tonnes, 40 million cubic metres, of waste is involved. This might not appear difficult, as the extractive industries (brick-making, gravel and mineral extraction, mining) produce holes amounting to no less than 250 million cubic metres – more than five times the volume of the annual accumulation of domestic waste. Unfortunately, many of these holes are unsuitable. They may be too far from the source of the waste, so that transporting it would not be economic: they may be too near to human habitations, and the well-known NIMBY ('Not In My Back Yard') syndrome means that there will be violent opposition to refuse disposal; or they may be close to sites of ecological importance, when conservation interests will object. Finally, and perhaps most important, the geology of the site may be unsuitable. All waste sites generate some leachate, produced when water seeps through their mass, and although this is unlikely to be highly toxic, it will include materials which should be contained and which certainly must not be allowed to escape into sources of water supply. For all these reasons it is often difficult to find sites for waste disposal.

The ideal landfill site is a large hole, covering an area of many hectares, in level ground in a clay soil. The thicker the layer of clay beneath the site, the better. Where a hole does not exist, an equally good, though much more expensive, solution is to erect a strong clay wall around a flat site producing what could be a lake if it were filled with water. In some cases it has been impossible to find convenient sites which are relatively impervious to liquids released from the refuse, and artificial liners made from bitumen, butyl rubber or various plastics (which are also used to line irrigation lagoons on farms) have been used. These have the disadvantage of having a limited life, so they can only be filled with disposable materials which do not emit soluble toxic substances for long periods.

In some instances, a hole excavated for mineral extraction can be used as a landfill site. Planning permission for this extraction is often given only if the exploiters undertake to restore the site to agricultural use. It may also be stipulated that the site should be filled with refuse, which is then covered with a thick layer of soil following compaction of the waste material. In such cases, the only likely problem is that if the hole is left for some years after the mineral extraction is finished, it may develop its own vegetation, become invaded by various animals, and (particularly if it is in an industrialized area) become of interest to conservationists. This may produce an outcry from wildlife enthusiasts, who do not wish the invading plants and animals to be buried under refuse. Such a situation is especially liable to occur with chalk excavations, as in parts of Essex near the Thames. In these somewhat unaesthetic conditions the best selection of orchids in the county is to be found.

Where a suitable hole in the ground does not exist, it is generally difficult to get planning permission to establish a landfill site. It is even harder to obtain the approval of the local inhabitants. Ideally the site should be near, but not too near, the centres of population which generate the refuse, it should have good access roads, and it should, if possible, not be on valuable (and expensive) land whose use is pre-empted by other organisations. The Ministry of Agriculture, Fisheries and Food stipulates that the best agricultural land, classified as grades 1 and 2 (which includes land on which most crops can be grown even in adverse conditions) should not be used, and the Nature Conservancy Council strongly opposes the use of any Site of Special Scientific Interest (SSSI), enforcing the provisions of the Wildlife and Countryside Act (1981) in support. Finally, the geology of the site should be such as to make it easy to contain any leachate that the refuse may generate.

There has been controversy and conflict between the local authorities and various conservation bodies regarding many proposed refuse disposal sites. In England it is the County Council that is responsible for the disposal of municipal waste. Councils have generally tried to find sites which will involve the rate-payers in the minimum expense. They have therefore often picked on areas of apparently derelict land, or, if it is farmed, areas of poor fertility (grades 4 and 5) where little opposition might be expected. As a rule they have taken conservation into consideration, and have avoided sites which have been scheduled as SSSIs. Unfortunately, in a number of cases, some sites approved for waste disposal have later been found to have value for wildlife, and have promptly been designated SSSIs by the NCC. The NCC's argument that it was not previously so described because staff shortages made it impossible to cover the whole country adequately has failed to convince the authorities

FINDING SITES FOR WASTE DISPOSAL   19

wishing to use the sites. They have complained that if the site was so valuable to be 'of international importance for conservation' it would surely have been discovered at an earlier date. This has given rise to much ill feeling, and the view that conservationists are deliberately trying to sabotage any kind of development.

I have personally experienced the difficulties in finding suitable sites, for my advice on the matter has been sought by various local authorities and commercial organisations. Sometimes it is clear that proposed sites do have considerable conservation value, and would best be left undeveloped, in which case the proposal has usually been dropped. On other occasions we have been able to find an alternative site to which there is less objection, and responsible conservation bodies like the NCC have accepted this fact. This has been true of some marshland sites bordering the Wash in Lincolnshire, where it proved possible to identify the 'least valuable' site, at the same time agreeing that other areas would be protected. A guarantee is also given that when any chosen site is filled, it will be restored so that it is at least of some value to wildlife.

Sometimes there may be strong differences of opinion about the possible effects of establishing a waste disposal site on or near an area of genuine wildlife value. If members of conservation organisations suspect that damage may be done to the site, they believe that their duty is to oppose it with all their resources. Unfortunately, they are not always entirely scrupulous in their opposition, and may greatly exaggerate the possible risk. The authority wishing to establish the waste disposal site is likely to recruit its own conservation 'expert' in support of its case. I have sometimes acted in this capacity and have been stigmatized by certain conservationists as a 'traitor' to the movement. I have no difficulty, however, in defending my position. If I thought the cause was entirely wrong, likely to do unacceptable damage to a valuable site, with no substantial compensation, I would refuse to appear on its behalf. If some damage was likely – say if 5 ha of a 100 ha marsh were to be lost, I would make this quite clear in my evidence; and I would endeavour to give a true picture of any possible wider effects on the rest of the site. Before giving evidence at any Public Inquiry, I would make sure that everything possible was done to restrict the potentially damaging effects of the development. And I would hope to obtain some sort of *quid pro quo* from the developer. Thus if one end of a marsh were lost, a larger area of a similar nature would be added at the other end.

I can illustrate what happened in a particular case. One of the biggest and most important waste disposal sites is at Pitsea in Essex. This has for many years taken an enormous amount of London's refuse, which reaches it by barges which sail down the Thames. The local authority, at that time dominated by extreme left-wing interests, wished to close down the site. Its ostensible reason was that as well as domestic refuse it also took chemical waste which they considered dangerous (see p. 55). The site was important to many industrial operators who had no other outlet for their waste. The way this chemical waste was dealt with, and its ecological effects, is dealt with in Chapter 5 on p. 55. The site could only reached from the landward side by a road which happened to belong to the local authority. The company operating the site wished to continue using this road, a solution which was supported by the Nature Conservancy. However, the local authority was adamant, and the company had to find another route. Fortunately it owned a considerable area of land, including some which connected with the main, public road leading to the site.

The only disadvantage was that a new road would have to pass across an area of reed marsh and open water known as Pitsea Hall Fleet, which was of agreed conservation interest and was part of an established SSSI. Therefore the NCC was bound to oppose this development, and in this it was supported by some, but not all, local conservationists. Much play was made by the opposition of the fact that the rare Bearded Tit or Bearded Reedling (*Panurus biarmicus*), of which at that time there were fewer than 400 pairs in Britain, nested there. The Inquiry was told, with every appearance of authority, that if the road were built the Bearded Tit was virtually certain to leave. One naturalist, opposing the road, said that 'the Fleet could become polluted with oil and rubbish'. The NCC representative stated that the road would 'pose a threat to the very existence of the marsh'. Another objector agreed that the marsh would cease to be a haven for wildlife.

In my evidence on behalf of the Company, I said I regretted that this road would have to be built, and that in doing so a small part of the reedbeds would undoubtedly be lost. I announced, however, that the Company had agreed to create a new and extended area of reeds at least twice that which would be lost – a feasible procedure, particularly as the reeds were naturally extending their coverage every year. I admitted that I could not be sure what would happen to all the species of birds in the marsh. Bearded Tits behave somewhat erratically, and several apparently satisfactory sites have, in the past, been abandoned for no known reason. However, I saw no reason why the road would harm all these birds, especially as parts of the marsh were more than 1 km from the proposed road.

The Inspector supported the application, and the road was built. Eight years after its construction the population of Bearded Tits had more than doubled; in 1986 one pair nested successfully within 10 m of the new road. The marsh has not become polluted, and all forms of wildlife which existed before the road was built are still present, generally in increased numbers. The site has, in fact, been improved, partly because the Company decided voluntarily to restrict the area used for waste disposal, and even to leave a large area of 'grazing marsh', which had been granted planning permission for tipping, uncovered and undeveloped. As grazing marshes are becoming increasingly rare, this was a substantial contribution to conservation in that part of Essex.

An interesting case of conflict occurred at Huntspill in Somerset. The County Council was having great difficulty in finding suitable waste disposal sites – no convenient 'holes' existed in the area. There was even a proposal to send refuse half-way across England to the brick pits in Buckinghamshire, a long and very expensive train journey. After studying all other possibilities, they picked on a flat area of 50 ha on rather poor farmland some distance inland from Bridgewater Bay. The conservation objection to this site was that parts of it were within 1 km of the important Bridgewater Bay National Nature Reserve, an international site for wildfowl registered under the International Ramsar Convention.

The NCC and other conservationist bodies claimed that tipping would affect the reserve in various ways. Vehicles bringing material would pass within 300 m of the shore, the boundary of the reserve, and would cause disturbance. The refuse would cause pollution of the land, of the sea and of nearby rivers. Most important, the refuse would encourage great numbers of birds, in particular gulls, which would prey on protected species within the reserve. The gulls would also be a nuisance in neighbouring towns.

The County Council answered these charges. Vehicles would be hidden from the sea by an embankment. The site would be so engineered that any leachate would be contained and purified. Finally the Council had devised a net (see page 30) to cover the site which would prevent it from being exploited by gulls and crows. These points were accepted by the Inspector and his Assessor, though the development was in fact turned down on other grounds, including aesthetic objections to a 'hump' in this flat coastal area, and possible noise affecting nearby homes.

There was one interesting development during these discussions. The NCC announced that it was considering declaring the whole of the proposed waste site as an SSSI, on the grounds that it was an important feeding and roosting zone for the wildfowl and wading birds using the reserve. This had never been previously suggested, certainly not prior to the waste disposal proposal. The NCC also supported its suggestion by describing the area as 'unimproved grassland', although it did not claim that the reserve contained many interesting plant species. There was indeed some grassland near to the shore, and the birds no doubt roosted there, but most of the waste disposal site was arable, and had been growing rather poor crops of cereals for some years.

In Wales there have been several cases where bogs and marshes have been proposed as waste disposal sites. There was considerable concern when Carmarthen District Council sought planning permission to deposit domestic refuse on Llanwllch Mire, a raised bog, already an SSSI, said 'to be storing 10,000 years of vegetational and agricultural history in its peat layers'. In this instance the ecological evidence was such that the Council was convinced and withdrew its proposal, the Public Inquiry was cancelled, and other sites were developed.

There are several interesting cases of man-made sites which have developed conservation value. In order to make it practicable to take waste to comparatively distant sites, transfer stations are being developed up and down the country. Here the refuse collected in urban dustcarts is tipped out and then compressed into bales, which spread no litter, and which take up the minimum space in lorries or railway trucks. After compression, the refuse is of little interest to birds, but while awaiting treatment at the transfer stations it attracts great numbers of gulls. A proposal to establish a transfer station in a disused quarry at Cowthick in Northamptonshire was opposed partly because gulls were likely to be a nuisance, but mainly because the original quarrying had exposed a unique complex of fossil submarine channels which permitted a reinterpretation of the environmental conditions under which mid-Jurassic Lincolnshire limestone was laid down. This is said to be a nationally important research and teaching resource. In this case plans were modified so that the most important geological sections have been preserved.

While I was writing this book, I learned of a case which might have been invented to illustrate the difficulties of waste disposal authorities and of conservationists. This was described on BBC TV *Country File* on Sunday, 4 December 1988. There is an abandoned quarry known as Cliff Rigg, near the village of Great Ayton, and immediately adjacent to the North Yorkshire Moors National Park. A proposal was made to restore the area to what it was before quarrying took place, using refuse as infill for this purpose. To the County Council, this seemed a scheme that should be generally approved. However, it was strongly opposed by conservationists and people living in Great Ayton. The latter said they would willingly pay higher rates for their refuse to be taken to a distant dump. The main objection, however, was that the quarry now had great ecological value, and also that it contained valuable geological exposures. It is perhaps ironic that the people who opposed restoring the quarry would create the greatest commotion were it now proposed to create another quarry in a similar site!

Chapter 3

MUNICIPAL WASTE SITES

In the past, refuse was simply dumped into holes. It blew about and littered much of the surrounding countryside. It attracted all sorts of vermin, and little attempt was made to control them. From time to time some soil was spread to bury the material, but this was generally done rather inefficiently. The dump stank and was an eyesore.

Today, practically every site managed by a local authority or responsible commercial firm well deserves to be called a sanitary landfill site.

Fig. 1 shows how such sites operate. They are divided into cells which are used in sequence. Trucks bring in refuse as collected in city streets during the first part of the day – this is often completed before 1 pm. In most large sites the material is then compressed by a special machine called a compactor. This resembles a large road roller with a very heavy spiked roller which breaks up large objects and produces a much denser material with fewer air spaces. The compressed refuse is then covered with some 10 cm of soil. Next day a further layer is added, and this continues until the cell is full. The process is repeated with other cells until the whole site is covered in refuse and soil.

Although the material is compacted, there is still a good deal of subsidence, and in large sites which take many years to fill this may allow further waste to be added. Eventually the level may be raised to that

Fig. 1 Sanitary landfill site in pit previously excavated for clay used in brickworks, now partially filled.

24 WASTE AND POLLUTION

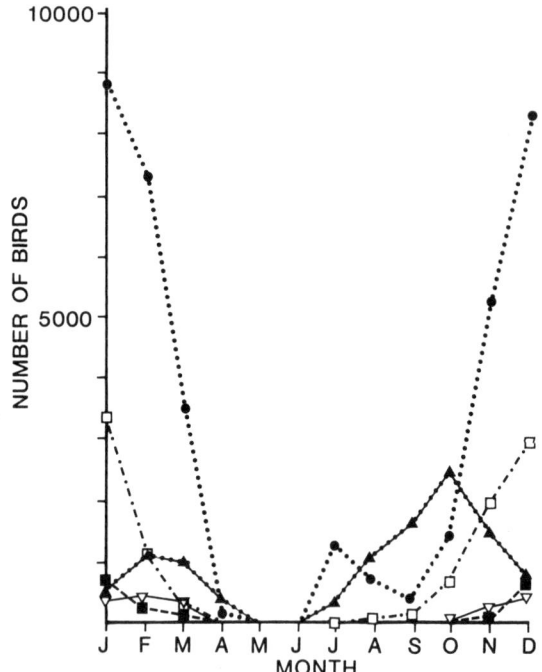

Fig. 2 Average monthly numbers of gulls of different species visiting refuse tips in south-east England, 1976-78. (After Horton, Brough and Rochard). Key: ●, black-headed gull; ❑, herring gull; ▲, lesser black-backed gull; ∆, common gull; ■, great black-backed gull

originally agreed on the site licence issued by the local authority for the operation. Today an increasing number of sites, instead of consisting merely of filled holes, are being developed into 'mountains' which not only accommodate much more refuse on the same area of ground, but can be merged into a new landscape in what is otherwise an uninteresting flat piece of country. In such situations, care is taken to cover the surface with a relatively impenetrable layer of clay, so that rain water does not penetrate into the mass of refuse and generate leachate.

While tipping is going on, perhaps for a period of many years, the site supports its particular forms of wildlife. The most important phenomenon is the invasion by birds of the area where the actual deposition is in progress. The modern, well-managed site only permits easy feeding for a short time from dumping until the refuse is buried with soil, and the constant passage of vehicles tends to drive off some species, but nevertheless many gulls are able to eat their fill in a few minutes. The larger gulls, such as the Greater Black-backed, are able to recover some of the food even when it is buried under quite a depth of soil.

The birds which most often visit refuse sites are gulls and corvids. Much the commonest gull is the Black-headed Gull (*Larus ridibundus*). Others are the Herring Gull (*Larus argentatus*), the Lesser black-backed Gull (*Larus fuscus*), the Common Gull (*Larus canus*) and the Great black-backed Gull (*Larus marinus*). Various other gulls, including some uncommon migrants, are often found in small numbers. Of the corvids, the Carrion Crow (*Corvus corone*), the Jackdaw (*Corvus monedula*) and the Rook (*Corvus frugilegus*) are almost always present. The Raven (*Corvus corax*), which was a widespread scavenger a hundred years ago, is found on refuse tips in Wales, and the Chough (*Pyrrhocorax pyrrhocorax*) occurs on some Irish tips. Other very common species usually present in considerable numbers are the House Sparrow (*Parus domesticus*), the Starling (*Sturnus vulgarus*) and the Feral Pigeon (*Columba livia*).

All these species feed on the freshly deposited refuse. They may be joined from time to time by the Pied Wagtail (*Motacilla alba*), which is more interested in insects, and frequently seen on the mature parts of the site. The Kestrel (*Falco tinnunculus*), which feeds mainly on small mammals on all parts of waste disposal sites, may be attracted by sparrows feeding on the recently deposited refuse.

Gulls have been studied more thoroughly than the other birds feeding on refuse. These birds have changed their habits during the last hundred years. Before that they deserved their common description of 'sea gulls', as they were seldom found away from the coasts. Today this has changed considerably, and gulls are increasingly found far inland. Colonies of breeding Black-headed Gulls frequent marshes well above sea level. Farmers everywhere encounter gulls following the plough, seeking insects and worms in the newly turned soil. Inland reservoirs provide overnight roosting sites, and the short turf of playing fields, golf courses and aerodromes teem with invertebrates. Sewage works, particularly the old types with settling lagoons (p. 102), might have been invented to encourage gulls, and refuse tips provide a new and abundant source of food.

Refuse disposal sites not only provide much easily acquired food, but also adjoin areas of bare ground covered with soil which has not yet developed its characteristic vegetation. This provides the ideal conditions for the gulls' favourite activity of 'loafing'. They stand about apparently doing nothing for most of the day, occasionally returning to any available refuse for food, before returning at night to their roosts, which may be on reservoirs up to 50 miles away, or on more traditional haunts near the coast.

Careful studies by a number of ornithological scientists indicate that

26 WASTE AND POLLUTION

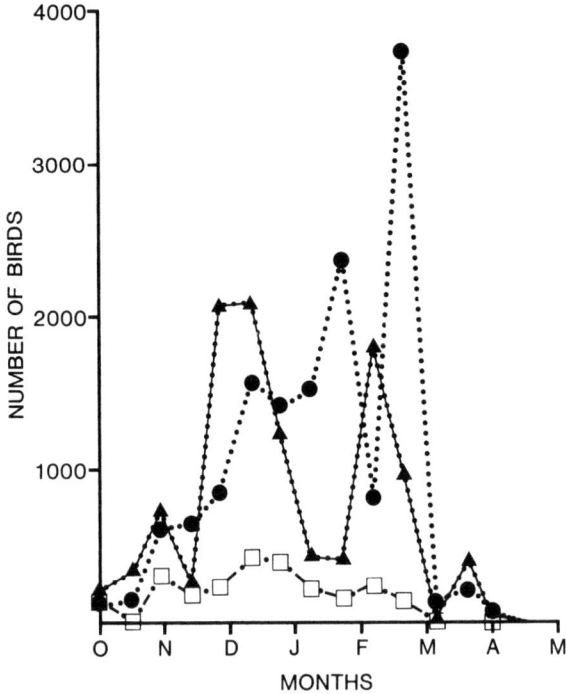

Fig. 3 Number of gulls at refuse tips in Worcestershire and Warwickshire 1977-78. Symbols as in Fig. 2. (After G.H. Green)

different species of gull feed on different sites at different times of year. The results of the work of N. Horton, T. Brough and J.B.A. Rochard of the Ministry of Agriculture Laboratory at Worplesdon are illustrated in Fig. 2. This shows that at five sites to the south-west of London, in the years 1976-78, the Black-headed Gull was the commonest visitor, and that it fed here mainly in winter. Four other gulls were present in lower numbers, but at approximately the same time of year. In May and June the gulls deserted the site, and fed elsewhere.

Somewhat different results were obtained in a study by G.H.Green of overwintering gulls on rubbish tips in Worcestershire and Warwickshire, virtually in the centre of England. The data obtained in 1977 and 1978 are shown in Fig. 3.

It is interesting to compare this with Fig. 2 for the London area. The Black-headed Gull was the commonest species in both cases, but in the second the Lesser Black-backed Gull was the close runner-up, and in several months gave the highest count. The Herring Gull, which came second in the first study, was a poor third in central England, and in this

MUNICIPAL WASTE SITES 27

Table 1 Gulls feeding at various sites adjacent to the Bristol Channel (after Mudge and Ferns)

Site	1		2		3		4	
	S	W	S	W	S	W	S	W
Refuse tips	68.9	75.4	48.2	42.5	16.3	47.2	0.2	5.3
Sewage outfalls	0.6	3.5	0.0	4.4	0.0	5.7	5.1	14.3
Littoral and inshore areas	6.2	5.2	20.0	5.2	50.5	47.2	1.1	11.8
Fields	16.0	15.3	15.8	46.4	0.5	0.0	93.4	67.1
Freshwater areas	6.4	0.1	12.0	1.0	9.3	0.0	0.1	1.0
Other areas	1.9	0.5	4.0	0.5	23.4	0.0	0.1	0.5

Key: 1, Herring Gull; 2, Lesser Black-backed Gull; 3, Great Black-backed Gull; 4, Common Gull. S, summer; W, winter.

Table 2 Average total population counts of gulls at feeding and roosting sites in south-east Wales in winter. The percentage at each site appears in brackets (after Mudge and Ferns).

Site	1	2	3	4	5
All roosts	20,925	744	23	4721	44,968
Refuse tips	15,779	316	15	7	2,394
	(75.4)	(42.5)	(47.5)	(0.2)	(5.3)
Sewage outfalls	738	18	1	241	5,506
	(3.5)	(2.4)	(5.7)	(5.1)	(12.3)
Littoral and inshore areas	1,080	39	15	52	5,299
	(5.2)	(5.2)	(47.2)	(1.1)	(11.8)
Fields	3,328	371	4421	31,769	
(calculated by difference)	(15.9)	(49.9)	(93.7)	(70.7)	

Key: 1, Herring Gull; 2, Lesser Black-backed Gull; 3, Great Black-backed Gull; 4, Common Gull; 5, Black-headed Gull

area only 13 Great Black-backed Gulls and 12 Common Gulls were captured. Other studies by J.C. Coulson showed the Herring Gull to be the commonest species in northern England, where Lesser Black-backed Gulls are virtually unknown.

G.P. Mudge and P.N. Ferns of University College, Cardiff made an interesting study of the feeding ecology of the same five species of gull on refuse sites adjacent to the Bristol Channel. They found that refuse was an important source of food in both summer and winter. Their results are summarized in tables 1 and 2.

These show that in both summer and winter Herring Gulls fed mainly at refuse tips, and that Lesser Black-backed Gulls came a good second. Black-headed Gulls, the commonest species in the London area and in central England, did not breed in the Bristol Channel area, and were almost absent during the breeding season. When present in winter, in greater numbers than any other species, they fed mainly on the fields, and hardly used refuse tips at all.

Mudge and Ferns investigated the importance of different food sources during the breeding season. This they did by handling large chicks from various gull colonies; the chicks spontaneously regurgitated their last meal. Herring Gull chicks had obtained nearly 70 per cent of their food from refuse tips. Many of the chicks studied were on the islands of Steep Holm and Flat Holm, well out in the Bristol Channel, and quite a few miles from the refuse tips where the parents collected the food. The results from Herring Gull chicks in nests on Cardiff rooftops were generally similar, but contained more material from freshwater sites in the vicinity.

Detailed long-term investigations by J.C. Coulson and his colleagues have concentrated on Herring Gulls in north-east England, where this species is the commonest gull to frequent refuse tips. Using marked birds, they found that these gulls appeared to be less dependent on refuse than the same species in other areas. Most birds feeding on the Coxhoe tip in County Durham did not do so on consecutive days – for the most part they paid a maximum of two visits in a week. Here 509 was the peak number of birds present at any one time during the day, 840 individuals visiting at intervals during the day, and 2,745 during the working week. This tip was well managed and the refuse was efficiently covered at the end of the week, so that little or no food was available. This was recognized by the gulls, few of which came there on Sunday.

Dr. Coulson's team also showed that Herring Gulls frequently carried some strain of *Salmonella*. These pathogens were also found in Black-headed Gulls and Lesser Black-backed Gulls. The *Salmonella* were probably picked up when the gulls were feeding on sewage, though decaying refuse might also have contributed. The gulls' faeces, deposited on ground grazed by sheep and cattle, were capable of contaminating the pasture. The gulls which were carrying the *Salmonella* did not seem to suffer any illness. It should be noted that although there is no 100 per cent proof that the gulls actually caused any outbreak in domestic stock, the circumstantial evidence is considerable.

Other important findings of this team were based on ringing and recapture of gulls in many European countries. Some 20 per cent of the Herring Gulls which winter in north-east England breed in Arctic Norway. The remainder nest mainly along the entire east coast of Scotland as far north as Orkney and Shetland, with only a few nesting locally. Wintering Black-headed Gulls mainly breed around the Baltic. Most Great Black-backed Gulls had bred in Norway.

These studies show that food provided directly or indirectly by man is

MUNICIPAL WASTE SITES 29

of major importance to gulls. This extra source of food partly explains the population explosion of these birds which has occurred in Britain. The increased numbers are also partially explained by some of the overflow from continental countries, where even greater numbers may have been produced, coming to Britain. The move inland may in part be due to increased numbers overcrowding the traditional sites nearer to the sea. Other factors are the new supplies of food, not only in refuse tips, but also in sewage works, and to some extent, the public habit of feeding gulls. The increased areas of mown grass in playing fields and aerodromes has clearly been significant. However, even with all these factors, gulls would probably not have travelled so far from the sea had there not been available increasing numbers of reservoirs and areas of open water in disused gravel pits where they could roost at night. Roosts up to 20 miles from feeding grounds are common, but many gulls fly longer distances twice every day. Thus gulls feeding in Cambridgeshire may roost on the River Stour nearly 50 miles away, so they fly 100 miles every day when this feeding pattern continues.

The size of the gulls' population explosion is difficult to quantify accurately. Locally it can be enormous. Thus in 1901 there were only 12 pairs of Herring Gulls on the island of Steep Holm in the Bristol Channel. In 1975 the population had increased to an estimated 14,484 pairs. The food from the waste disposal sites must have played a part in this phenomenon.

This altered situation has not been generally welcomed. Gulls feeding on refuse have often carried away lumps of decaying food, and dropped them in built-up areas or on agricultural land. This has added something particularly offensive to the already objectionable litter. It is also a health hazard, for various pathogenic micro-organisms have been recovered from the dropped food, as from the birds' excreta. Some gulls, particularly Great Black-backed Gulls, prey on the chicks of other species. In the studies of Mudge and Ferns, Herring Gull chicks were most commonly found being fed to the nestlings of Great Black-backed Gulls, but it is well known that many shore-nesting species, including rarities such as occur in the Bridgewater Bay National Nature Reserve, are at risk. The gulls nesting on town roofs are particularly unpopular, as they are both smelly and noisy.

Various attempts have been made to prevent gulls and crows feeding on refuse sites. Men have been employed to scare them away, using guns with loud, blank cartridges (public opinion, despite objections to the birds, would be aroused if they were actually shot), broadcasting recordings of repellent noises, and even launching falcons which may, in fact, make the odd kill. On one site these measures were estimated to have reduced the gull population from 11,000 to under 1,000, but if the bird-scarer takes a day off the gulls quickly return in undiminished numbers. However, if scaring continues long enough it may significantly affect the population by removing one important food source.

The most successful method now in use is the Mobile Cover Net, invented and developed by Mr Jon Henton, an engineer working in the County Surveyor's Department in Somerset. This device has been patented, and is coming into increasing use in other parts of Britain and on the European continent. It received a Commendation in the 1983 Pollution Abatement Technology Awards made by the Confederation of British Industry, the Department of the Environment and the Royal Society of Arts.

The problem was to produce a net which would exclude large birds from the area where refuse was being deposited. It would have to be large enough to accommodate more than one refuse truck and a compactor which could compress the material deposited; and it would have to be easily moved from one part of the site to another. Furthermore, its supports would need to be strong enough to prevent it from collapsing or overturning in strong winds.

The final version is a box-shaped net, 36 m long and 36 m wide, carried on eight 12 m masts. Each mast is carried on a trolley unit, designed not to overturn even in a full gale. Each trolley unit comprises a high-strength reinforced concrete frame mounted on four reinforced concrete roller wheels. The frame measures 5 x 3 m. Originally 10 m wooden masts were used, but in the more recent modification 12.5 m steel masts have been substituted. The trolley units are connected by chains, and the four on one side of the net can be moved easily by one or, if some distance is involved, two tractors. The net itself has a mesh size of 85 cm, which keeps out gulls and corvids, retains virtually any wind-borne litter, but allows small birds to pass through easily without being trapped.

The net is certainly very efficient. It was first erected during a weekend when there was no tipping. When work started on the Monday morning the usual population of some 2,000 gulls and some hundreds of corvids arrived in anticipation of an easy meal. During the morning, when only a few birds usually feed, some puzzled birds circled the net and settled down to loaf until lunch time. When the men departed, the whole mass of birds took off and flew towards the area where they hoped to feed. Some flew into the net, but most took avoiding action. They continued their noisy flight, and showed every sign of extreme frustration. Many of the birds continued their fruitless attempts to enter the feeding area, even during the afternoon when work had resumed. In the end they gave up and flew off to their roosting places.

On Tuesday, the normal population of birds returned to the site, repeating the previous day's performance. After that they apparently learned their lesson, for much smaller numbers came on Wednesday, and by the end of the week the site was practically deserted. As long as the net was in position, the birds did not come back. However, as soon as it was removed the gulls returned in their original numbers within three days, giving the impression that 'scouts' had been keeping the tip under observation, and that when the food was again available they had passed on the information.

The net was first tried out at the Saltlands site at Bridgewater in Somerset, where it worked well in winds exceeding 100 km per hour. The staff particularly appreciated the way it retained the litter, and cut down the unpleasant task of collecting it. Nearby residents were also pleased when waste paper and plastic no longer blew into their gardens. The net has been equally successful at Preseli in West Wales and at Portree in the Isle of Skye. In West Germany, where winds are notoriously high in the region of the German Bight, litter control has been very successful. The trade name of the appliance is 'Somernet', and its current cost is about £50,000.

It is clear that we now have a technique which should make it possible to operate a landfill site without damaging the amenity, and the wildlife, of the surrounding area. Some useful 'holes in the ground' left by the extractive industries near to towns or National Parks may now be safely employed for waste disposal.

So far we have considered what happens during the period when active tipping is under way. As soon as the refuse is properly buried, it starts to decay. However, the soil cover is not always complete, and some gulls and other birds may, for a time, continue to feed. Rats will also burrow into the capping soil, and therefore will not be restricted to the places where dumping is in progress. In fact, rats may be found all over mature sites, living on plant material in the same way as they exist widely throughout the countryside. Insects, including house flies and blow flies of various species, may breed and be reinforced by individuals brought in with fresh refuse (although they seem to be fewer than when covering and burying the refuse was less efficiently done). Soon after it is deposited the refuse generally starts to heat up, as the putrescent material begins to decompose; conditions here are similar to those obtaining in a garden compost heap. The hottest spots may be temporarily unfavourable to most organisms, but the outer zones may be warm enough to support animals adapted to hotter climates than that of Britain.

Fires often break out during this period. These may be started spontaneously by heat generated within the refuse, by smouldering material tipped on to the site, by deliberate vandalism or accidentally by careless workers dropping lighted matches. Fires will obviously damage wildlife, and temporarily prevent revegetation, but little permanent harm is likely.

One interesting exotic insect found in refuse sites is the domestic Cricket (*Gryllus domesticus*). This is a native of dry regions in southwest Asia and North Africa, which became established in British bakehouses and other warm buildings where food was plentiful. Houses in which fires burned all through the winter were also infested, and the singing of the 'cricket on the hearth' was a familiar sound in Victorian kitchens. Decomposing refuse tips provided ideal conditions, and some developed enormous cricket populations. Other exotic insect species, again originating in north Africa, which invaded warm houses in the 16th century were the German Cockroach (*Blatella germanica*) and the Common Cockroach (*Blatta orientalis*). Although all these insects can still be found on some refuse sites, the majority which I have studied in recent years are no longer infested, mainly because of changes in site management. Also they will die out during cold winters when the decomposition ceases and the temperature falls.

In the early months after tipping, even with compaction, air is able to diffuse into the mass of material, and so aerobic decomposition proceeds. Rain percolates into the refuse, and unless it is excessive the moisture also encourages the breakdown of the food residues and other putrescible material. This produces considerable heat, and discourages plant growth where the surface dries out. Carbon dioxide and hydrogen sulphide will be produced, and these gases will reach levels below the surface which will discourage mammals from burrowing and plant roots from penetrating. Inflammable mixtures of gases are not generated at this stage. The aerobic decomposition of plant and animal material will be virtually complete in six months, at the end of which period the temperature will fall. After this the long-term changes within the refuse will start to occur.

A great concern of those managing refuse sites is that they may generate leachate, produced when rain, surface or ground water percolates through the contents of the site. Although leachate is seldom highly toxic, as heavy metals and other poisonous substances tend to be when firmly bound on to the material deposited, it usually contains considerable amounts of salts and organic matter. For this reason, leachate must not be discharged untreated into rivers and watercourses, although it can be used to irrigate certain grasses and other plants. In a site with an impervious lining, of clay or of some plastic material, it may be possible to contain the leachate, at least for some years. In the drier parts of eastern England, this should theoretically be possible indefinitely, for the total potential evaporation over the whole year is generally substantially greater than the rainfall. In winter, when most rain may fall, evaporation is minimal, and rain fills up pools and saturates the soil. In summer evaporation is much greater, and it may be necessary to add many centimetres of irrigation water to keep the soil in equilibrium. Much the same happens in an outdoor swimming pool. In winter the water level will rise as rain falls. In summer, in eastern England, notwithstanding further rain, the water level will fall, with a net fall of perhaps 20 cm during the year. In the wetter west of England, there is more rain and less evaporation so over the year water levels may rise.

A site with an impervious lining in eastern England might be thought to generate no leachate, with the refuse drying out as a result of evaporation. Unfortunately this does not happen. Even compressed refuse contains voids, and water runs through, producing a wet layer at the bottom which his not in contact with the surface. So when the surface dries up, much of the water in the lowest part of the site will remain there, and in time heavy rain in winter may overflow the edge of the site and cause pollution problems.

Site operators take great pains to prevent this happening. In dry areas in summer the water lying at the bottom is pumped up and sprayed over the surface of the refuse. It then evaporates away. This system has an additional advantage for the leachate sprayed over the soil irrigates it, and encourages the growth of grass and other vegetation. Although most leachate is not poisonous, some plants react unfavourably to the high salt content, and much research, particularly in the United States, has been carried out to find species of grass which grow best under these circumstances, and which play a part in removing potentially damaging substances from the leachate. For a small site adjacent to a large river or to the sea, it may be possible for the water authorities to grant a licence to the site operator to discharge leachate into the water near by. This is only allowed when it can be shown that the amount of salts and organic matter in the leachate will not upset the ecology of the water which receives it. Where this is impossible, treatment plant is erected to purify the leachate and to remove all substances that might have detrimental effects when discharged.

While any part of a site is in operation, rain will penetrate and produce leachate. When tipping ceases, the site must be developed so that leachate ceases to be a problem. Care must be taken in contouring the site, to avoid hollows which will collect water. Also the site must be capped with clay or other material not easily penetrated by water, so that the rain will run off harmlessly and not make contact with the refuse within the mass of the site.

Fifty years ago domestic refuse contained over 50 per cent coal ash, and much less paper and cardboard than it does today. It contained considerable amounts of waste food and vegetable matter, which decayed quickly, producing heat and carbon dioxide in the manner already described. At the end of six months, or a year at most, the refuse was more or less stabilized and the site was safe to develop, perhaps for housing.

The situation today is very different, as many local authorities and building contractors have found to their cost. When the aerobic decomposition is complete, conditions within the mass of refuse change, and anaerobic decomposition begins to take place. The carbon in the paper and cardboard, which now makes up a third or more of the material, is transformed very slowly into methane gas. Methane or 'Firedamp' may now continue to be generated for as long as 20 years. It is not inherently poisonous, is lighter than air, and will diffuse upwards out of the refuse. This is the main constituent of North Sea gas. Previously coal gas was used in British homes. Since it consisted mainly of highly poisonous carbon monoxide, it was frequently used for suicide attempts.

Although the methane produced by refuse is not poisonous either to plants or animals, it does discourage revegetation of some sites, as its presence in the soil prevents oxygen from penetrating, so that roots are asphyxiated. Burrowing animals are also excluded by the lack of oxygen. The greatest disadvantage of methane is its danger to humans, particularly if houses are erected on a site where it is still being generated. Under such conditions, explosive mixtures, which may contain as little as 5 per cent methane, may be produced. Should these be generated into a house with an unventilated cellar, serious explosions may occur. As a result of several serious accidents, greater precautions are now taken, and many potential building sites may be idle for up to 20 years, while the owners wait for anaerobic decomposition to be complete, resulting in the end of the generation of methane.

The long period of anaerobic decomposition also has ecological effects. Heat is still generated for many years, in sufficient amounts to be detectable by infra-red aerial photographs. This heat tends to dry out the surface and to damage newly planted trees. Moreover, until decomposition is complete, the surface level will not be completely stabilized, so buildings erected too soon could be damaged, and tree roots could be broken.

The situation is sometimes mitigated by inserting ventilation shafts into the decaying waste, so as to accelerate the discharge of methane. Gas-proof membranes are included in the foundations of houses to prevent the methane from entering them. (The same technique is used where underlying rocks generate radioactive radon gas, p. 153.) But these palliatives are not entirely successful, and it is no exaggeration to say that under present-day conditions, it is cellulose and not any highly toxic chemical which is the most dangerous substance in our municipal waste.

This is one good reason why recycling of paper should be made more attractive. At present it is seldom economic, particularly with paid collectors of waste paper. However, potential losses must be tiny compared with the costs of not building on what would otherwise be an ideal site for much-needed houses.

In some cases the generation of methane has been turned to financial advantage. On certain sites, particularly those which are deep, with a restricted surface area, the methane can be collected and used as a fuel. According to the XIth report of the Royal Commission on Environmental Pollution, there are some 25 large landfill sites in Britain which merit development as fuel sources. Already methane is collected and burned on a considerable scale from a number of sites. There is, incidentally, one environmental benefit in burning methane, instead of allowing it to diffuse freely into the atmosphere. Methane is the second most important 'greenhouse gas' (see p. 133), and when burned, produces carbon dioxide. Although carbon dioxide itself is the most important greenhouse gas, methane, volume for volume, has a much greater effect, so that burning it should make some small contribution to the reduction of global warming. While large, shallow sites producing a limited amount of methane over a large area cannot readily be exploited as sources of this fuel, its possible use should perhaps be taken more into account when new sites are planned. A smaller, deeper site will be easier to exploit, and the value of the methane might well be included in the calculations when deciding between two sites with different configurations. Incidentally it is not always realized that if paper and card are efficiently recycled so that they are not included in refuse, the economic production of methane will cease.

Most studies of the ecology of refuse tips, including the book with this title published by Arnold Darlington in 1969, are concerned with sites where the flora and fauna have developed after tipping is complete. There are, of course, many large and important sites which are still in use, while other parts of the area have not seen fresh refuse for many years. Such sites may, from the ecological point of view, be the most important as they have the potentiality to develop the greatest diversity of wildlife.

In the past a number of refuse tips, when filled to capacity, were given the minimum cover of soil, and left without further systematic restoration. In time they have developed the sort of vegetation which springs up naturally on any neglected site. The fact that they are full of decayed refuse and not natural soil seems to make comparatively little difference. First they are covered with aggressive weed species, then by shrubs such as hawthorn and wild rose, the seeds of which are readily carried there by birds. Within a few years tree species, including oak, ash and sycamore, appear, and a 'natural' wood is produced. This may be quite pleasant to look at, adding to the local amenities, but few of the characteristic herbaceous species of old, natural woodland will appear spontaneously, although if they are deliberately introduced many of the features of ancient woodland may be produced. It is true that with insufficient cover some of the waste materials, old prams and rusty tin cans, may come to the surface, sometimes exposed by burrowing rabbits or even by dogs playing on the site, and there may be a predominance of nettles, docks and other species which flourish in soils with a high nitrogen content, derived from the underlying refuse. But in time the wood may settle down and support a characteristic population of birds and insects which, being mobile, soon colonize such places.

Nowadays the licence to establish a waste disposal site usually contains conditions as to its after-use, and this affects the way in which the surface is treated. Thus if it is to be returned to agriculture, a much deeper layer of soil may be needed as compared with a site to be grassed and used as an amenity, such as a playing field. In Britain the normal specification for most sites is a cover of one metre of topsoil, preferably over a further layer of subsoil. In many cases this may be excessive, and an unnecessary use of the limited (and expensive) amounts of topsoil which are available. Furthermore, work on the revegetation of colliery spoil heaps has shown that, with the careful use of fertilizers, or with the spread of filter cake (that is, sewage sludge, partially dried) grass may be established equally well with subsoil alone. However, it is important to determine the stage of decomposition of the refuse to be covered. If the anaerobic breakdown and the consequent generation of methane has ceased, then a much shallower soil cover is necessary, compared with the conditions of active anaerobic decay when the layers of refuse presents conditions which are lethal to plants, so that their roots cannot penetrate to any considerable depth.

When a site is to be returned to farming, the type of farming should be specified. If it is proposed to grow arable crops like wheat, which send their roots down several metres into the ground, only a mature site with a deep cover of soil will be suitable. If the site is to be grazed by sheep, good grass can be grown on quite a shallow covering of soil, particularly if the necessary fertilizers are applied as required (as a rule, little and often). Where the grass is grazed sufficiently, but not too heavily, the top layer of soil will be modified and improved. But with today's refuse it is seldom possible to restore a waste site to grow good cereal crops in less than 20 years.

Sites have been used as 'amenity grassland' quite soon after being filled. The continuing production of methane does little harm, as on an open area dangerous levels do not build up, and the gas is diluted and blown away by the wind. Sites may be developed for nature conservation, or planted as public parks. Experiments to determine which species of trees are most likely to grow successfully on a recently filled site have proved somewhat surprising. On one site, trials by the Forestry Commission showed that the pioneer species of natural communities such as poplar, willow, alder and birch, which are frequently recommended for landfill sites, did not do at all well, making poor growth and frequently dying in the first two years after planting. It was perhaps surprising to find that the species which proved most promising were Holm Oak (*Quercus ilex*), Field Maple (*Acer campestre*), Small-leaved Lime (*Tilia cordata*) and Ash (*Fraxinus excelsior*). Sycamore (*Acer pseudoplatanus*), an aggressive alien tree which is spreading rapidly in many old woods to the detriment of native species, does not do well initially, but those which survive may grow well. Hawthorn (*Crataegus monogyna*) and Guelder Rose (*Viburnum lantana*) were among the most successful shrubs, as was also Tamarisk (*Tamarix* spp), a species which flourishes in salty seaside areas, perhaps indicating that salt levels are also elevated on refuse tips.

On many landfills, where no development is expected for many years, to allow time for the decomposition to proceed, areas of bare soil are left, and here spontaneous revegetation takes place more or less rapidly. Observations by P.J. Shaw on the large Pitsea landfill site in Essex illustrate the process in a region with a low rainfall; ground cover might have been more rapid in the wetter west of England. This site is particularly well managed, and the newly deposited refuse is covered with soil each day when tipping is finished. During this observation the covering soil was mainly clay, and it appeared to contain few seeds or other propagules, so the plant cover was mostly derived from plant species growing on adjacent parts of the landfill.

Table 3 shows how areas where the refuse was covered with soil in different years developed its cover of vegetation. Thus in 1981 the soil cover established in 1977 had a 91.69 cover, so there was little bare soil left. The patch covered in 1978 had, in 1981, a 75.02 cover. However, the 1980 cover, after one year, had only established an 18.22 cover. At each site described here, the composition of the vegetation was sampled in ten permanent quadrats by recording the interception of needles

38 WASTE AND POLLUTION

Table 3 Mean per cent vegetation cover [n=10] on newly deposited soil, Pitsea landfill site

Year	Vegetation cover during 1981	Vegetation cover during 1982
1981		46.77
1980	18.22	76.26
1979	45.55	81.37
1978	75.02	87.74
1977	91.69	87.30

Table 4 Comparison of species abundance (mean percentage cover) during the first year of succession, i.e. site 1980 during 1981; site 1981 during 1982. (nr = not recorded in this survey but present on site; np = not present at site)

Species	Mean Cover Site 1980	Mean Cover Site 1981
Creeping Bent, *Agrostis stolonifera*		1.66
Slender Foxtail, *Alopecurus myosuroides*		5.97
Oat Grass, *Arrhenatherum alatius*	nr	1.03
Orache, *Atriplex prostrata*	2.23	
Common Wild Oat, *Avena fatus*	np	2.63
Rye Grass, *Lolium perenne*	2.43	9.40
Timothy Grass, *Phleum pratense*		1.06
Annual Meadow Grass, *Poa annua*	3.91	3.12
Meadow Grass, *Poa pratense*		3.00
Rough Meadow Grass, *Poa trivialis*	3.37	1.91
Common Knotgrass, *Polygonum aviculare*	3.26	7.63
Creeping Buttercup, *Ranunculus repens*	nr	1.31

attached to thin steel rods. Data were recorded at ten points in 35 randomly generated positions per quadrat. Cover is used here in the sense of non-repetitive vegetation cover.

The same table shows results obtained using the same methods a year later, in 1982. It will be seen that the same general pattern was followed, but that there were considerable variations. Thus in the studies made in 1982 the vegetation developing during the first year was considerably greater than in 1981. The differences of cover in the older plots are not statistically significant.

Table 4 shows the main species of plants making up the cover in the first year of succession of the two plots at the top of the previous table, i.e. of the 1980 plot in 1981, and the 1981 plot in 1982. It will be seen that, of the 12 species concerned, nine were different grasses. The other three were common 'weeds' of farmland.

Table 5 repeats, in column 1, the data given in table 2, and then, in the next column, shows how vegetation developed in the next year. Several

Table 5 Percent species contribution in site covered in 1980

Species	Mean cover during 1981	Mean cover during 1982
Creepng Bent		37.17
Orache	2.23	nr
Creeping Thistle, *Cirsium arvense*		1.97
Lyme Grass, *Dalymus repens*		1.69
Rye Grass	2.43	8.20
Ribbed Melilot, *Melilotus officinalis*	nr	1.60
Timothy Grass	nr	1.94
Bristly Ox-tongue, *Picris echioides*		9.57
Annual Meadow Grass	3.91	
Rough meadow Grass	3.37	
Common Knotgrass	3.26	
White Clover, *Trifolium repens*	nr	2.53
Red Clover, *Trifolium pratense*	nr	1.14
Creeping Buttercup	nr	1.77
Curled Dock, *Rumex crispus*		1.37

Table 6 Percent species contribution in site covered in 1977

Species	Mean cover during 1981	Mean cover during 1982
Yarrow, *Achillea millifolium*		2.34
Creeping Bent	22.08	10.20
Oat Grass	2.66	1.20
Mugwort, *Artemesia vulgaris*	1.11	2.31
Creeping Thistle	1.40	2.94
Spear Thistle, *Cirsium vulgare*	1.17	
Cocksfoot, *Dactylis glomerata*	1.14	5.80
Wild Carrot, *Daucus carrota*	1.54	2.26
Common Teasel, *Dipsacus fullonum*	1.69	2.63
Lyme Grass	5.83	28.94
Yorkshire Fog, *Holcus lanatus*	15.97	4.34
Ox-eye Daisy, *Leucanthemum vulgare*	1.17	
Rye Grass	8.94	10.03
Timothy Grass		1.23
Bristly Ox-tongue	16.63	3.51
Ribwort Plantain, *Plantago lanceolata*	?	?
Creeping Buttercup	?	?

new species, including Creeping Thistle, Bristly Ox-tongue, and Red and White Clover made their appearance. Bristly Ox-tongue was the second most successful species, after Creeping Bent which covered nearly 40 per cent of the area studied.

Table 6 shows the results of rather long-term colonization. The site covered with soil in 1977 was examined four years (1981) and five years (1982) later.

Considerable changes, as compared with the studies of years 1 and 2, had occurred. Only four of the nine original colonizing grasses remained, and only Creeping Bent and Rye Grass continued to have a substantial (over 10 per cent) cover. Lyme Grass had appeared, and in 1982 covered the most ground (29 per cent). Yorkshire Fog was also relatively successful. Two of the broad leaved 'weeds' (Orache and Common Knotgrass) had disappeared. Creeping Buttercup remained, and had been joined by Mugwort, Creeping Thistle, Spear Thistle, Wild Carrot, Common Teasel, Ox-eye Daisy, Bristly Ox-tongue and Ribwort Plantain.

Appendix 1 lists 134 species of flowering plants, including shrubs, which have been found on the Pitsea landfill site. I suspect that this list is still incomplete, and that over 200 species may eventually be discovered. This compares very favourably with many Nature Reserves and Sites of Special Scientific Interest.

We have rather little systematic information about the way in which animals colonize a waste disposal site. There are many records of flying insects and of birds which have been found on such sites, and examples are given in the appendices. These animals are mobile, and presumably find their way to these sites in a more or less random manner, so a casual sighting is of little significance. It is only when the vegetation has developed so as to provide the right species of leaf for a caterpillar, or seeds for a species of bird, that these wanderers will remain more or less permanently.

Any large landfill site is likely to support populations of most of our small mammals. Rats and mice are particularly attracted by the refuse, but are also found well away from areas of fresh tipping. When the vegetation has started to grow, and when insects are becoming common, the Wood Mouse (*Apodemus sylvaticus*), the Field Vole (*Microtus agrestis*) and the Common Shrew (*Sorex araneus*) soon build up substantial populations. These animals have been studied to see whether they build up raised concentrations of heavy metals or other toxic substances which may be contained in the waste, or which may be added to the refuse if the site is used to co-disposal of other wastes (see p. 61). In most cases these small mammals did not seem to be affected by any such substances which were present in the refuse or in the soil used to bury it.

Any large landfill site is likely to support a colony of Rabbits, (*Oryctolagus cuniculus*); numbers are generally controlled by outbreaks of myxomatosis, with a build-up when the epidemic comes to an end. Hares (*Lepus capensis*) are likely to visit the site as soon as any areas of grass have developed. Weasels (*Mustela nivalis*) seem to be early colonizers, perhaps attracted by the numerous mice and voles. Quite a number of sites support healthy Badger (*Meles meles*) colonies. Sometimes setts are dug into an area of old waste, and the burrows seem to run through masses of plastic and household appliances with little soil. The badgers do not seem to mind this unaesthetic mess, and to flourish among it. Foxes *Vulpes vulpes*) appear to be frequent visitors, seeking the abundant food supply, rather than permanent residents. I have not myself observed a Fox's earth on a waste site. Hedgehogs (*Erinaceus europaeus*) are commonly found, but not, in my experience, Moles (*Talpa europaea*).

Landfill sites do not as a rule provide suitable roosting facilities for bats, but the large number of insects which may be found in the air above them has been found to attract various species. The Pipistrelle (*Pipistrellus pipistrellus*), as would be expected, is the species most commonly present, but the late Lord Cranbrook regularly observed the Noctule bat (*Nyctalus noctula*) and the Serotine Bat (*Eptesicus serotinus*) at a tip in Suffolk; and there are reliable records of Daubenton's bat (*Myotis daubentoni*) and Leisler's bat (*Nyctalus leisleri*) at a tip on the Herts-Essex border. Perhaps if bat boxes were installed near a landfill site, even more species might be encouraged to live in their vicinity.

I have already mentioned the birds which feed on newly tipped refuse. Later on, as the site matures, it is likely to be visited by very large numbers of birds. Appendix 2 lists the species seen at the Pitsea Marsh between January 1971 and April 1987. The landfill site is central to this area, but is also includes reedbeds, open water (the Pitsea Hall Fleet), grazing marsh and thorn scrub; so although the birds were not found exclusively on the landfill area, it is probable that most if not all visited it, and some actually bred there. These data are given here as indicating how a refuse disposal site may form a useful part of a complex area which is very rich in wildlife.

Wildfowl, including several duck and geese, were often seen on this landfill, and in fact wildfowlers greatly appreciated the privilege of shooting them there. For several years the wildfowlers also established breeding pens on an area where tipping had previously occurred, and raised considerable numbers of Mallard (*Anas platyrhynchus*) and various geese.

At all stages of its development, Kestrels (*Falco tinnunculus*) are likely to be seen hunting on any landfill site. As there has been considerable public concern about the risk of pollution from such sites, especially where they are used for co-disposal of toxic wastes, I thought that it would be interesting to use these kestrels to elucidate the situation. Kestrels, which live mainly on small mammals, but which also eat some birds, are considered to be at the end of the 'food chain'. In this case they may, under suitable circumstances, concentrate poisonous substances present in lower and apparently harmless amounts in their prey. There are many well documented instances of predatory birds being damaged in such situations.

Although kestrels were often seen visiting and hunting at the Pitsea landfill site, there were no large trees or other objects where they could nest and lay their eggs. I remembered seeing nesting boxes used on the new Dutch polders, where a plague of voles had damaged the newly sown grass. The kestrels had been encouraged, in the hope that they would control the voles. In the event, vole numbers fell as the kestrels came to live on the polders, but there is some doubt as to how important they were in reducing vole numbers.

At Pitsea we installed a number of nest boxes on the Dutch plan, with the object of seeing whether the kestrels, which would feed mainly on mice and voles on the landfill site, would flourish or whether they would be harmed by any substances found there. In fact, for several years the majority of the nest boxes have been occupied, normal eggs have been laid and hatched, and a considerable number of the young have developed and flown away. Unfortunately, many of the young have been stolen from the nests, probably by people who saw the film *Kes* and who hope to train the birds as 'falcons'. But the experiment has at least demonstrated that there are no significant concentrations of poisons passing along the food chain, as might have been feared if the site were not so carefully managed.

My main conclusion is that a waste disposal site, whether or not it is used for the co-disposal of toxic wastes, has great possibilities for the support of wildlife. This should be borne in mind by those who automatically oppose any development of such sites, as they may actually improve the capability of an area to support a rich and varied flora and fauna.

Chapter 4

MINING WASTES

The Mole (*Talpa europea*) has been described as the first miner. Mole hills are composed of its waste, the spoil derived from its tunnellings underground being voided on to the surface. Mole hills have a considerable ecological effect locally. In some upland areas in Wales mole hills may appear to cover up to half the area of some pasture, thus considerably reducing its value for grazing. The freshly broken-up soil also provides sites for germinating seeds, and in this way many weeds invade well-developed pasture of mainly economic grass species. The mole hills may even become sites for invasion by hawthorn and other shrubs, as the sheep or cattle tend to avoid the hills when grazing, although hawthorn seedlings in really well-grazed grass seldom survive. They also bring worms and soil insects to the surface, and one way of detecting whether moles are active underground is to observe when blackbirds and thrushes approach the hills where new soil and fresh food appear. It is difficult to estimate the total weight of the mole's mining waste, but throughout Britain it is likely to amount to at least a million tonnes in a year, so it can hardly be described as a trivial problem.

As shown in Chapter 1, however, man's efforts in this direction are very much greater, as mineral workings and industry between them produce, annually, nearly 250 million tonnes of different wastes. Some of these, as those produced in stone quarries, are valuable raw materials easily absorbed in road building, but much waste is dumped in massive heaps, disfiguring the landscape. Today some of this is being removed, and many heaps are being imaginatively vegetated, in ways described by A.D. Bradshaw and M.J. Chadwick in their outstanding book *The Restoration of Land*. Restoration is usually of a utilitarian nature, for amenity or for some form of recreation, but this could be (and occasionally is) amended to produce some form of vegetation valuable for wildlife conservation. It is perhaps ironical that there has recently been some local opposition to the restoration or removal of large spoil heaps. Coal tips, china clay deposits and even hillsides covered with slabs of broken slate from quarries are sometimes considered to be part of our 'industrial heritage' and therefore candidates for preservation in their starkest form. It is perhaps fortunate, though the process may be very slow, that they are likely eventually to develop some concealing vegetation, and they may in the very long term become covered by shrubs and trees. But, in most cases, some process to accelerate vegetation is clearly desirable.

Coal mining

Considerable areas of ground are required when deep coal mines are established. In Britain in recent years this has given rise to controversy in such places as the Vale of Belvoir, which is quite a distance from earlier mines. In this case the objection has mainly been to the change in a pleasant agricultural landscape, but wildlife will clearly be affected, particularly by the spoil which will have to be excavated in addition to the coal. In some mines the environmental damage done by the spoil is only temporary, for there are plans to put it back underground. As this is costly, however, it is by no means a universal practice, and as a rule the spoil is dumped in small mountains all around the area of the pit.

These characteristic black 'mountains' are a particularly familiar sight in County Durham, parts of Yorkshire, Nottinghamshire, South Wales and several areas in Scotland. In the past they were generally left untouched, and for many years almost no plants appeared on the surface. Seeds often germinated, and growth started, but the plant suffered the fate of the seed that fell 'on stony ground' in the Parable of the Sower. The spoil heaps were often steep-sided, and therefore unstable, and small 'landslides' were common. Very occasionally these gave rise to disasters such as occurred at Aberfan in South Wales on 21 October 1966 when 144 individuals including 116 school children were fatally buried by a collapsing spoil heap.

Plant growth is difficult on coal mine spoil, because of the nature of the material. It consists almost entirely of finely crushed rock and coal, with little or no soil and almost no nutrients. Furthermore, it often contains pyrite (iron sulphide) in quite large quantities. This discourages growth, by producing very acid conditions which may be between pH 2 and 4. The acid material, and the associated metal, may also be leached out of the spoil and may pollute neighbouring watercourses. In one such case an unpolluted stream, with slightly alkaline water (pH 7.9) and very low concentrations of metal, received leachate from a mine. Below the spot where this occurred the pH remained in the region of 3.3 for over 10 km, and iron levels were raised to 40 ppm. Effects of such water pollution are considered in more detail in Chapter 5.

In recent years, there have been many very successful attempts to vegetate coal spoil. This requires treatment of the surface layer with lime – possibly as much as 50 tonnes per ha in very pyritic soil – and fertiliser. In some cases deep layers of expensive topsoil have been put over the surface, in other cases only subsoil has been used. The addition of any kind of soil makes restoration easier, but is not strictly necessary. The lime is generally well retained, and has a more or less permanent effect so far as it penetrates. Inorganic fertilisers need to be applied lightly and often because heavy dressings before the vegetation is established tend to be washed away. As well as soil, organic dressings such as sewage sludge may be applied successfully. The trouble here is that sewage is liable to contain high levels of metals, particularly lead, which may make the ground unsuitable for food crops or gardens where vegetables for human consumption are grown. If only amenity grass is required, sewage can generally be used. Actually, even fairly high levels of lead are seldom taken up by animals living on grass grown on such soils, and wild mammals generally show no sign of being affected.

Trees planted directly into coal tips die, even if fertilised. If planted after a grass cover has been established, they may survive, although they will only grow slowly. The best results have been obtained by digging holes and filling these with a couple of wheelbarrows full of soil. Such trees will grow quite well at first, but will only continue to flourish if the surrounding spoil has been limed and continues to be fertilised for some years after restoration begins.

Whereas well-limed and fertilised tips will grow most grasses used in agriculture and found on rough grazing areas in Britain, sometimes the best results are obtained, with minimum trouble and expense, with strains of grasses which are tolerant of high levels of any existing heavy metals. Collieries in Britain differ considerably from one another. Some tips are reasonably low in metals, others contain large quantities. It is always wise to analyse the spoil at any site where revegetation is planned. In most cases, it is essential to introduce some permanent drainage system, particularly as the lower layers may be very impervious.

Open-cast or strip mining is the cheapest, quickest and most economical method of winning coal from the earth. It is widely practised in many countries, particularly in the USA and Australia, where the environmental impact is often horrendous. This was at one time also true of Britain. Today open-cast mining is governed by stringent planning regulations. It is normally only allowed where it will help in the restoration of land already mined, although some lower grade agricultural land has also been used. As a result of these restrictions, open-cast mining in Britain has little effect on areas of conservation interest or on wildlife. It is permissible to dig up only areas of minor importance, and it is also always

46 WASTE AND POLLUTION

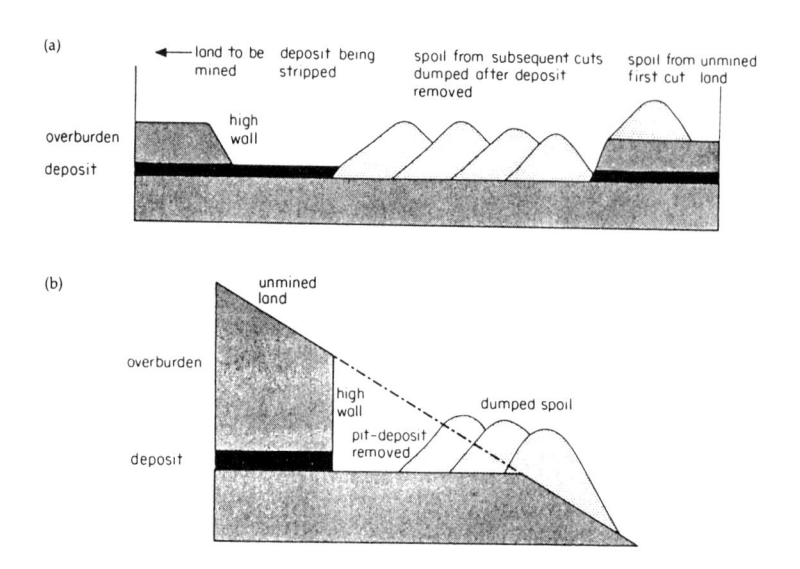

Fig. 4 Open cast mining: a, Shallow deposit in flat country; b, In hilly country. (After Bradshaw and Chadwick)

stipulated that the land should be left in an improved condition, generally for farming.

Figure 4 shows how open-cast mining operates. There is usually only a comparatively thin deposit of coal, although it must be thick enough to be economically worth excavating. To get at it, the overburden of 'ordinary' soil and rock has to be removed – this overburden is generally 20 to 30 times the volume of the coal. A hole is dug down to the coal layer, this is removed, and the first lot of spoil is dumped on to the surface. This is the only serious excursion outside the excavated area. The rest of the spoil is piled within the excavation after the coal has been removed. In some cases the topsoil for the whole area to be mined is ripped off and stored, but this may be an unnecessary complication, as the topsoil soon deteriorates unless spread comparatively thinly, and useful soil fauna like earthworms do not generally survive. When restoration is to take place, after all the coal has been removed, the spoil is spread out evenly over the cavity. Although the volume of the spoil is less (by the volume of the coal) than that of the material removed, it usually fills the hole even when compacted as much as possible; the subsidence in succeeding years is generally minimal. When such sites have been efficiently drained and appropriately fertilised, they have often been restored to full agricultural productivity. The depth of soil has proved sufficient to grow good crops of cereals. Where hedges and trees have been planted, many former open-cast sites are difficult to distinguish from neighbouring fields which have not been so exploited. Incidentally, on open-cast restoration areas, as elsewhere, any trees planted will grow best if grass is discouraged for a circle of up to a metre all around, so allowing them to obtain better supplies of water and fertiliser. Successful rehabilitation of open-cast areas depends on good planning and careful after-use, and proper application of fertiliser is crucial. Some of the best results have been obtained where pasture with a good complement of clover has been grazed for a number of years before returning the fields to arable cultivation. As mentioned above, if required for wildlife conservation, such restored land has great potentialities, but may be too expensive to be acquired for such purposes.

Iron ore

The world's richest iron ore deposits are in Australia, and other important sources exist in America. However, iron ore has been mined for many years in Britain, and some of the environmental effects still remain. Strata of iron ore, generally much thicker than those of coal, are found quite near to the surface. In the early days of mining, the ore was removed with minimum damage to the countryside. The overburden was removed in wheelbarrows, the topsoil generally being kept separate. Comparatively small areas were excavated at a time, so that even with few mechanical aids the topsoil was only stored for a brief period. When the ore had been removed, the subsoil and then the topsoil was replaced, and the field system, some depth below the original surface, was restored. As a rule, the fields were immediately able to yield reasonable crops.

At the beginning of the 20th century, 'progress' struck the industry in England. Immense steam-powered excavators were employed, profitably in Leicestershire and Northamptonshire, to remove iron ore at greater depths than had been previously exploited, so that a much greater amount of overburden was removed. As this was a period of agricultural depression, there was little economic inducement to rehabilitate the land. The mining companies could buy agricultural land with underlying iron ore so cheaply that they were prepared to abandon it at the end of their operations. The result was that, by 1945, there were over a thousand hectares of totally derelict land in what was otherwise a rich agricultural region. The overburden had been replaced with the least possible trouble, mixing any topsoil with the mass of the material, and the result was 'hill and dale' land, which was difficult even for walking.

Much was simply neglected. The soil was comparatively lacking in nutrients, particularly phosphorus, but nevertheless some vegetation started to appear. The main species found included Couch Grass (*Agropyron repens*), Bush Grass (*Calamagrostis epigejos*), Creeping Thistle (*Cirsium arvense*), Spear Thistle (*Cirsium vulgare*), Beaked Hawksbeard (*Crepis vesicaria*) and Rosebay Willow Herb (*Chamaenerion angustifolium*). Seedlings of shrubs including Hawthorn (*Crataegus oxycanthoides*) and Goat Willow (*Salix caprea*) and trees such as Silver Birch (*Betula verrucosa*) and Oak (*Quercus robur*) appeared but seldom survived. Legumes, which might have done much to improve the levels of nitrogen, were generally absent, presumably because of the low phosphorus levels. The growth was generally so poor that it could not even be said to produce some semblance of a 'wilderness', and the sites did not prove attractive to birds or small mammals. This was presumably in part due to the poverty of the soil fauna – earthworms in particular were virtually absent.

The only remotely economic use made of the unimproved hill and dale land was to plant trees. Hand planting of 2-3 year transplants, especially of Larch (*Larix decidua*) and Sycamore (*Acer pseudoplatanus*) was carried out. Where fertiliser, particularly phosphate, was added, the growth has been reasonable, and quite respectable 'woods', with all the appropriate wildlife, have developed. Even the soil fauna has slowly returned. These areas, with appropriate management, would be particularly suited to wildlife conservation, as the forestry, although aesthetically more pleasing than the derelict ground, is not very productive.

Since the Second World War planning regulations have insisted on full restoration of ironstone workings. In 1951 the Ironstone Restoration Fund was set up, with a levy based on every tonne of ore extracted, to cover the costs of restoration. The sites are carefully levelled, usually (as in the 19th century) lower than originally, but sometimes extra material has been brought in to restore the original ground level. The area is cross-ripped, and any large stones near the surface are removed. Disking and harrowing to produce a satisfactory tilth then follows, fertiliser (based on soil analysis) is applied, and a good grass ley generally results. It has been found that these soils need substantially more fertiliser for some ten years than do normal agricultural soils, and these extra applications are paid for by the Ironstone Restoration Fund. At the end of ten years the restored land is hard to distinguish from the surrounding countryside where no mining has taken place.

In the past, ironstone mining has destroyed areas of ancient seminatural wood and other features of conservation importance. Today such developments would not be permitted. The restored land, like that derived from open-cast coal mining, could clearly be used for many purposes in addition to agriculture. It could be returned to forest, and this might be desirable in situations where mining areas are still adjacent to ancient woodland areas, from which emigration of plants and animals would be possible. Under these circumstances, something very similar to genuine ancient woodland should be produced.

Sand and gravel

The building industry has an almost insatiable appetite for sand and gravel. In Britain some 100 million tonnes of these materials are dug from 1,000 ha of generally good agricultural land every year. Although there is very little actual waste from sand and gravel digging, some holes made by the extraction may be used as waste tips, and the gravel workings often provide valuable wildlife sites.

Many deposits of sand and gravel are in river valleys, and the deposits are below the level of the water-table. Thus when the material is removed, the holes fill with water. The layer underlying the excavations usually has a comparatively high clay or silt fraction, and therefore is a satisfactory medium for plant growth. The nutrient levels are frequently adequate for reasonable plant growth. So, just left to itself, a wet gravel pit soon develops its characteristic vegetation, and requires comparatively little management, particularly in the early stages. If good beds of reed (*Phragmites australis*) are required, some judicious planting may be useful, and may accelerate the process, but many species, including Reed Mace (*Typha angustifolia*), Yellow Iris (*Iris pseudacorus*), Great Willow Herb (*Epilobium hirsutum*) and Water Mint (*Mentha aquatica*) will soon grow on the margins, and various pond weeds in the open water. Before long Alder (*Alnus glutinosa*), Willow (*Salix alba*) and Sallow *(Salix cinerea*) may appear and flourish on the banks surrounding the water.

Almost anywhere in Britain where open areas of water are created, duck and other birds move in. Mallard (*Anas platyrynchos*), Moorhen (*Gallinula chloropus*) and Coot (*Fulica atra*) are almost inevitable arrivals, and many less common species will almost certainly accompany them. The growth of gravel pits in the last century has allowed the Great Crested Grebe (*Podiceps cristatus*), which was an endangered species a century ago with only 32 pairs in England, to increase to a healthy population of well over 5,000 pairs. Gravel pits are also quickly taken over by many aquatic insects and other invertebrates, and fish, including Perch (*Perca fluviatilis*) and Roach (*Rutilus rutilus*) flourish – it is often uncertain how they find their way into these locations.

While almost all wet gravel pits are potential wildlife habitats, only a proportion remain available for this purpose. Some are used for noisy sports such as water skiing, which discourage the animals, others are restored to agricultural and recreational use. This is not easy, as suitable filling material may be hard to find. As the water communicates with that outside, domestic waste, which would cause pollution, is unsuitable, although it has been used when the material is sealed off by a thick butyl rubber lining. As such a lining is unlikely to last for many years, only refuse which is relatively free from toxic substances and organic matter can safely be used. Also, particularly when used for recreation, the areas of open water may have a higher cash value than the same area of agricultural land. Thus in many parts of England, e.g. the Cotswolds and the valley of the Great Ouse, the whole face of the countryside is being changed, with increasingly large areas of open water being created as the gravel is dug out and transported to building sites. Those areas employed for boating and other sports will encourage some forms of wildlife, even where they are far from being sanctuaries, so the contribution to conservation is generally positive.

Dry gravel pits are usually on good agricultural land, grade 2 at least, and there is generally pressure to restore them to the farmers who previously worked the land. In such areas the topsoil is normally preserved while the gravel is removed. These pits can be filled with relatively inert materials, such as pulverised fly ash from power stations (see p. 51 below), which is then covered with at least a metre depth of subsoil and topsoil. With suitable management, and the correct application of fertilisers, the land is soon brought back almost to the state it was in before gravel extraction began.

Clay for brickmaking

All over Britain where there were deposits of clay which, when baked, produced a hard mass, brickworks sprang up. Certain small enterprises produced very attractive products, some of which continue to be made and have a luxury value. However, most small brickworks have closed, and their sites may be used for other purposes. One at Ramsey Heights in Huntingdonshire has been made into a centre for the study of wildlife by the Bedfordshire and Huntingdonshire Wildlife Trust. There are similar developments in other parts of the country.

Today almost all the major brickworks are in Buckinghamshire, Bedfordshire and around Peterborough, on the strip of Oxford clay which runs diagonally across England. The works are all immediately adjacent to the pits from which the clay is dug. The clay is near to the surface, so only a small amount of surface soil has to be scraped off. After that, only brickmaking clay is dug out, work being stopped as soon as the best material is exhausted. As a result there is virtually no waste to dispose of.

Although the material at the bottom of the pit may not be top-class brickmaking clay, it is still relatively impervious material, and the pits are ideal places for the safe disposal of waste. Many are used for this purpose, and trains and lorries bring a continual stream of municipal waste to Buckinghamshire from London and other industrial towns. The sites are well managed, and the waste is always quickly covered with soil, although in the past these operations were not always so efficiently performed. The intention is to restore all brick pits for agricultural, recreational or building purposes. As mentioned above, p. 34, where domestic refuse containing much cellulose is used, the continuing generation of methane makes it dangerous to build houses, but grass and even crops have been successfully grown.

One very large enterprise has been the use of the brick pits near Peterborough to contain pulverised fuel ash (PFA). Modern, coal-fired electricity generating stations utilise coal which has been ground to a fine powder and is then blown in an airstream into the furnace. The carbon is burned off, and produces the heat which generates the power. The residual ash melts to produce PFA, which is made up of tiny, colourless, glassy spherical particles. Some 95 per cent of PFA is, in size, similar to sand and silt. It contains only 1 per cent of clay-sized particles, but nevertheless it has fairly good water-holding properties. It contains no organic matter at all, may be very alkaline (pH 11), and is relatively free from nitrogenous salts, but otherwise it contains at least some of the more important plant nutrients. PFA by itself tends to set hard, like cement.

When this type of power generation was first introduced, it was thought that disposal of the PFA would be a major problem, but in fact it has been less serious than was expected. Nearly half the PFA has been used to produce building blocks which are light and have good insulating properties. In fact, it has been suggested that efforts should be made to utilise more of the PFA for this purpose, instead of digging holes in the ground in order to extract clay to make bricks, and then filling up the holes with the PFA which could itself have produced the 'bricks'. However, the London Brick Company, which operated the Peterborough Brick Works, entered into a contract with the Central Electricity Generating Board, to allow the CEGB to deposit their PFA and then to restore the land before returning it to the brick company.

The process called for a number of tanker trains to convey the PFA, which contained enough water to make a slurry that could be piped into the brick pits. These were then filled to near the surface level required. The advantage of the PFA slurry over some wastes used to fill holes was that it did not contract after deposition, so there was no risk of subsidence. Then the PFA surface received one of several possible treatments. When only some 5-10 cm of sewage sludge or other organic matter was worked into the surface of the PFA, reasonable grass growth was possible. Poor crops of wheat could be obtained under the same circumstances, if nitrogenous fertilisers were applied. The best results were obtained with a deeper layer of soil worked into the top of the PFA and fertilised. Fairly good tree growth was obtained when planting was effected in pits 30 cm in diameter and 40 cm deep, filled with good soil.

One difficulty with the PFA is that, when fresh, it contains a quantity of the metal boron, which is toxic to many plants, particularly such crops as barley, peas and beans. If the PFA is allowed to weather before planting, much of the boron is leached out. Boron levels also fall considerably during the first years of any sort of cultivation, so it does not seem to present a long-term problem.

At Peterborough, the restored brick pits filled with PFA have been used to grow various crops, and as recreational grass. They have been proposed as sites for house-building, but some authorities are not yet convinced that the ground will be sufficiently stable. However, recent tests have been reassuring, and it is likely that building will in fact go ahead.

The brick pits, whether in course of excavation, or after filling with PFA, do not appear likely locations for wildlife conservation. But at Peterborough part of the area of the old pits has been turned into a nature reserve. One pit has not been filled, but has been allowed to collect rainwater to produce a lake. Trees have sprung up or been planted on areas between excavations, and a management plan has been devised with wildlife conservation as its main objective. The reserve is still in an early stage, and its flora and fauna will in time be enriched, but it is already proving a valuable educational facility.

China clay

Kaolin or china clay is mined in Devon and Cornwall, and has gone far to destroy nearly 100 sq km of the countryside. As well as being used in the British fine porcelain industry, and the manufacture of paper, it is also a major export. Kaolin deposits were produced by the hydrothermal degradation of granite from below, extending to great depths, so digging continues deeper and deeper, and the waste cannot be put back into the pits. The amount of waste is considerable – one tonne of kaolin is

MINING WASTES

accompanied by 1 tonne of mica and 7 tonnes of mainly quartz sand. The sand is built up into mountains, the mica goes into lagoons in nearly valleys. The result, until recently, was a veritable moonscape. Natural revegetation does take place, but very slowly, and may be overwhelmed if more material is used to cover existing tips. One large heap, after 30 years, had a growth of grasses and shrubs tolerant of the existing acidity on its lower slopes, but the rest was still virtually uncovered. Where old clay working becomes colonised by leguminous species, thus raising the nitrogen levels, trees may appear and grow reasonably well, but this process is likely to take at least 100 years. The industry only began in 1770 (previously rougher clays were used by the pottery industry, and kaolin was chosen to compete with the fine porcelain from China, hence 'china clay') and some of the oldest heaps are today still not completely covered with dense vegetation.

Recent work, again mainly by A.D. Bradshaw and his colleagues, has shown that the process of vegetation can be greatly accelerated. It has been usual to leave the steep mountains of spoil, as with plant cover these do not get badly eroded by the weather. The structure of the sand mountains is poor, and they would dry out badly in the east of England, but in the much wetter west country this seldom occurs. Fertilisers and lime must be applied, and hydroseeding (i.e. spraying the seed mixed into a jet of water) is generally used on the steeper slopes. A suitable seeding mixture has been devised, containing Rye Grass (*Lolium perenne*) to establish a quick cover, and Red Fescue (*Festuca rubra*) and Common Bent (*Agrostis tenuis*), appropriate local grasses, to build up a permanent cover. Legumes, Red Clover (*Trifolium pratense*), White Clover (*T. repens*) and Bird's-foot Trefoil (*Lotus corniculatus*), needing adequate lime and phosphate, are required to accumulate nitrogen in the sward. Tree lupins (*Lupinus arboreus*) and gorse (*Ulex europaeus*) grow well and increase the build-up of nitrogen. Eventually a wide selection of trees can be introduced.

The mica waste is usually deposited in ponds. If these are drained good 'soil' results. This retains fertilisers well, and can be used to grow good crops of vegetables such as potatoes. The mica waste may be utilised for cropping the same year as that in which it is produced. Obviously many types of plant can be grown in it.

As the original 'moonscape' is, in most areas, left intact but covered with a good vegetation of herbs and trees, the site will always be unsuitable for normal agriculture or ball games requiring flat land. It has therefore obvious potential for wildlife management and for education.

Quarries

Immense quantities of rock and stone are today extracted from quarries for use as building materials, for roadmaking, and for the production of cement and lime. In the modern quarry there is little waste, as almost all the rock can be utilised in one way or another. In some old quarries the overburden remains as a mound which is easily covered with vegetation; but in most cases it is used to cover the scars produced by quarrying.

There is generally very vocal opposition to opening up any new quarry, particularly when it is in a National Park or area of scenic beauty. Sometimes the people who are most vocal in their condemnation of the electrical industry for polluting the air with sulphur dioxide also lead the opposition to quarrying for the limestone necessary for operating the flue gas desulphurisation (FGD) which removes the gas.

The slate industry of North Wales has produced the greatest amount of waste, and today there are heaps containing perhaps 500 million tonnes in the areas where quarrying was important. These quarries produced slates for roofing, in various sizes, and for tombstones and memorial plaques. The accuracy involved in this work inevitably resulted in large amounts of useless offcuts. The mounds of this acidic rock are well described by Bradshaw and Chadwick as 'a miserable environment for plants'. A few mosses and the occasional birch may be all that relieves their surface. Fortunately, many slate heaps are now being removed for use in roadbuilding. There is then little loss of any wildlife habitat.

Although there is increasing pressure to reuse waste from limestone and chalk quarrying, any such material left over is rapidly colonised by a rich vegetation, and mammals and birds find shelter among the stones. Indeed, the main interest in quarries today is that they so often turn into valuable wildlife habitats. B.N.K. Davis has described what has happened in a series of chalk and limestone quarries, some of which have become the best sites for orchids and other interesting plants in their respective counties. This development of rare vegetation in quarries, particularly those in the chalk, or in limestone country, can be a source of embarrassment to the authorities. Some examples of this are quoted in Chapter 2. There have been many instances where a quarry has been excavated on the understanding that the site will be restored to its original contours and to use for agriculture. Where restoration has been delayed, the quarry may have developed into a site of conservation importance, and may even have been scheduled as an SSSI. As a result, there has been strong opposition on environmental grounds to carrying out the original plans for restoration.

Chapter 5

HAZARDOUS WASTES

The public, and the media, are very much concerned about the danger to themselves and to the environment from toxic wastes. This was the reason for the excitement generated by the *Karin B*, a ship loaded with leaking barrels of unknown substances, which tried unsuccessfully to land its cargo in Britain in the summer of 1988. Waste cargoes are in fact landed and processed in this country, which has caused us to be stigmatised in some quarters as 'the dustbin of Europe'. In the *Daily Telegraph* of 3 November 1988 we read: 'Grave concern at the Government's "very lax" attitude to the safe disposal of dangerous chemicals and industrial waste was expressed by Sir Hugh Rossi, chairman of the Commons Select Committee on the Environment.' We need to know how real this danger is, and what damage is done by hazardous or toxic waste to the environment and to wildlife.

There is no clear definition as to what is meant by 'hazardous waste'. The size of the problem may be inferred from the eleventh report of the Royal Commission on Environmental Pollution which estimated the annual production of hazardous wastes in England and Wales at 4.4 million tonnes. Some types of waste are covered by legislation. The Deposit of Poisonous Waste Act 1972 was one of the speediest pieces of legislation ever introduced in Britain. Before 1972 there were few regulations and, as will be shown, poisonous substances were deposited in many sites with little control and few special precautions. The 1972 Act was passed within days when drums containing cyanide were found on waste ground used as a children's playground. The Act introduced the concept of 'notifiable' waste, which was a blanket category comprising all toxic or dangerous waste which was not specifically excluded. It became an offence 'to deposit on land poisonous, noxious or polluted waste in circumstances which could give rise to an environmental hazard'. The Act also made it obligatory on those disposing of waste to notify the appropriate authority of the composition, quantity and ultimate destination of any notifiable waste.

The 1972 Act was effective, but cumbersome to operate because of the paper work involved. The Control of Pollution Act 1974 (COPA) was an attempt to deal with pollution and waste disposal in a comprehensive way. Under this Act, Control of Pollution (Special Waste) Regulations 1980 were introduced, and the 1972 Act was eventually repealed. The new regulations introduced a new category of 'special waste' to cover the more dangerous substances. These included medical products available only on prescription, or listed substances in dangerous concentrations. These are described, in somewhat macabre language, as having:

1. The ability to be likely to cause death or serious damage to a tissue if a single dose of not more than 5 cc were to be ingested by a child of 20 kg bodyweight, or
2. The ability to be likely to cause damage to human tissue by inhalation, skin contact or eye contact on exposure to the substance for 15 minutes or less, or
3. A flash point of 21°C or less.

The various categories of wastes are detailed in Appendix 3. It will be seen that the situation is far from simple.

However great the improvement after the passage of the Deposit of Poisonous Waste Act 1972, the country was, at that time, left with the legacy of the period when such regulations did not exist. The situation was not quite as bad as some people imagined, because since 1947 planning controls over all developments had been enforced, so that no new large-scale uncontrolled dumps had been established. Nevertheless there were many sites where all manner of toxic substances had been deposited during the previous hundred years, and little was known of the effects on the environment. The situation was studied by D.C. Wilson, E.T. Smith and K.W. Pearce of the Hazardous Materials Service of the Harwell Laboratory, and their findings were published in 1981.

The results of this survey were generally reassuring. The scientists looked into open dumps which had been abandoned in what might well be a dangerous condition, landfill sites which had been partially restored, and land contaminated by past industrial use.

Perhaps the most horrendous example of an open dump was at Malkins Bank in Cheshire, described as 30 ha of 'aesthetic insult'. This dump was developed in the late 19th century and used until 1932 as a salt and chemical works, pumping brine from the ground (where salt deposits were present) and using water from the stream for manufacturing processes. This left the stream culverted with a considerable area of the site overlain with lime sludge in three large lagoons. When manufacturing ceased in 1932, indiscriminate tipping of toxic and noxious waste from several different industries covered some 15 ha of the site to a depth of up to 15 m.

When examined in 1970, the tip was squalid and what was described as 'the odour of rotten eggs and tomcats' could be smelt miles away. Thousands of drums, containing a wide variety of chemical wastes, including reactive substances such as metallic sodium, lay in the sludge. There were bales of paper and stinking tannery wastes. Where tipping had not occurred, the lime sludge had dried out. The whole site was hazardous. Minor explosions were common, and there were open shafts, falling into which would almost certainly have proved fatal.

What might have been a very dangerous situation was greatly mitigated by the presence of a deep layer of lime sludge which covered most of the site, something which was also true of other sites in north-west England. This lime sludge, which neutralised many of the toxic materials, was used to bury the drums, which when possible were crushed before burial. The stream was seriously polluted, though much of this resulted from developments upstream of Malkins Bank. The stream was culverted to avoid further pollution. After levelling, the whole area was buried in at least 0.5 m of soil brought in from outside, and it was then grassed and developed as a golf course. Considering what remains underground, the site will have no further development of the kind likely to bring toxic material up to the surface, but as grassland it should offer no problems. The lime sludge and other materials will, in time, become incorporated in the soil. Obviously this site, used as a golf course, will have wildlife potential, particularly if this is borne in mind when the course is laid out.

Another site investigated by the Harwell team was on flat marshy land on an estuary subject to flooding at high tide. Tipping took place between 1945 and 1972. Large quantities of asbestos-cement, rubble and flux glasses were dumped along the river bank, in an attempt to protect the site from the river. Heaps of multicoloured chemical wastes, and a considerable amount of domestic waste, raised the fertility so that lush vegetation, including apple and other fruit trees, had grown up spontaneously. Although described as a 'moonscape', the site was used by motorcyclists and fruit-pickers. There is no suggestion that eating the fruit did any damage, although the soil proved to be seriously contaminated with asbestos fibre and heavy metals. The total for zinc, copper and lead reached 37 per cent, and the recorded maximum for cadmium was 0.5 per cent. This site has been reclaimed as a public open space, precautions being taken to minimise pollution of the adjacent estuary. It has been covered by a deep layer of soil, amounting to at least 1.5 m. Sadly, the productive fruit trees have been removed.

A third site had been used to deposit calcium sulphide waste from the old Leblanc process for the production of calcium carbonate. There are many such dumps in Lancashire and, as described below (p. 64), some have developed wonderful wild flower gardens. Others, where wet uncompacted refuse is also present, present problems. Any disturbance releases hydrogen sulphide, a very poisonous gas, smelling of rotten eggs when dilute, but generally undetected when in stronger and more dangerous concentrations. The pH here was 12-14. This site was buried deeply and grassed over successfully.

The most widespread British problem has been that of contaminated land arising from industrial practices which were normal a hundred years ago, or from waste disposal. A notorious case is the Lower Swansea Valley, where Victorian metal smelters have left their legacy.* In almost every town the old town-gas works has left its area of sites contaminated with coal tars, phenols and spent iron oxide (used for gas cleaning) containing up to 50 per cent free sulphur, together with sulphide, sulphate and various cyanide compounds. In the worst cases, the highly contaminated soil has been excavated and reinterred in a safe site. Otherwise the material has been buried. As a rule, care has been taken to avoid developments which would release the material from its otherwise safe location.

The most serious problems in the UK have been caused by the release of methane in housing estates (see p. 34 above). There have been cases where gardens of houses built on former tips contain unacceptable levels of lead, arsenic, cadmium, copper or zinc. If grassed over, there would probably be no danger, but the soil could contaminate vegetables and might be ingested by small children who are notoriously prone to ingest 'dirt'. Where such contaminated gardens have been identified, the soil has been removed and replaced.

Although we must continue to be vigilant and on the look-out for dangerous old dumps, and although we have no reason to be complacent about our success in dealing with this problem, conditions in Britain seem to be much better than those which obtain in the United States of America. The USA has a far greater area of land available, much of it well away from inhabited zones. In thousands of such places dangerous wastes have been dumped without any remedial treatment, so the dumps are still very dangerous. The most notorious is that at Love Canal, near Niagara. During the 1940s and 1950s, a chemical company disposed of

---

* Those who despair of having a clean environment with developing industry should note that the massive modern smelters, even when badly operated, only liberate a minute fraction of the pollution produced by the much smaller Victorian industries.

HAZARDOUS WASTES 59

its industrial wastes by sealing them in drums and burying them on this site, as was perfectly legal and according to accepted practice. The drums were then covered with soil, and a great many houses were erected on top of the waste. At first all went well, but then the drums corroded and the contents leaked into nearby water and percolated to the ground. On 7 August 1978, a generation after the area had ceased to be used for chemical waste disposal, conditions were so bad that Love Canal was declared a disaster area, and 239 families living close by were evacuated from their homes. Later, many more – almost the total population – had to be moved. There was obviously great local anger, and concern that people had been seriously injured or even killed by the pollution. The most careful investigations in fact showed that no one had suffered serious harm, but this was due more to good luck than to careful management. It is perhaps not surprising that some extreme environmental groups refuse to accept the official findings, and still insist that deaths and serious illness occurred.

According to the Eleventh Report of the Royal Commission on Environmental Pollution, there are in the USA possibly as many as 20,000 abandoned hazardous waste disposal sites and a 'Superfund' has been set up to clean up those for whom no responsible organisation can be traced. The Commission says: 'There is no evidence that in the UK waste disposal sites have, either in the past, or at present, been operated in a way likely to cause similar problems – certainly not on such a scale – and the need for emergency action is therefore much less likely. This is not to suggest, however, that there have been no examples of bad practice.' In the UK, the emphasis has been in the importance of reusing land for economic and social reasons, because 'new' land for all developments is scarce, particularly in and around urban areas. Such developments generate the funds to clear up the pollution. Also, notwithstanding the comments of Sir Hugh Rossi reported above, we do now have sufficient legislation to prevent further abuse of the environment, and the technical skills to put the regulations into force.

The question we have to ask is what were the effects on the environment, and on wildlife, of the uncontrolled dumping which went on before 1972, and which was even less controlled before the planning acts were passed in 1947. Clearly 'moonscapes' like Malkins Bank were virtually sterile, and the same was largely true of heavily contaminated land such as existed around gas works and in the Lower Swansea Valley. Some of the less heavily polluted soils have acted as outdoor laboratories, becoming spontaneously covered by vegetation. In many cases, as when the soil contains so much copper that the growth of normal commercial strains of Rye Grass is inhibited, the process of natural selection appears to have operated rapidly, and plants which are naturally resistant to copper have been able to grow and to flourish. These resistant strains, when identified, have proved very useful for growing on similar sites where selection has not occurred. Observations made on these polluted sites have, in fact, yielded a whole series of resistant species for such purposes.

Where toxic dumps have polluted rivers, the results have been devastating, as is described in Chapter 6. Many animal deaths must have taken place when poisonous materials were dumped, or when animals visited the contaminated sites. However, although there may have been local effects, there is little evidence that toxic wastes have had any serious effects on the national population of any forms of British wildlife.

Britain, in fact, has probably gained by being a small, densely populated country. Sites for waste have never been easy to obtain, and large, responsible companies have done much to reduce the waste problem. Ideally, a chemical manufacturer should have no toxic waste to dispose of; in many cases the amount that leaves the factory is quite small. Valuable substances like gold and silver, which might be included in 'waste', are almost completely recovered and reused – this form of recycling makes economic sense. Moreover, many poisonous substances may be treated within the works and rendered harmless, and today this is happening increasingly. Where a firm has developed the means of detoxifying and destroying its own poisons, it may use the same process to treat other people's waste, often providing the basis of a very profitable industry.

The fuss that was made about the importation of drums of polychlor biphenyls (PCBs) from Canada into Britain in 1989 illustrates the conflicts which may arise. PCBs, used widely in industry until the nineteen seventies, have been shown to be longlasting toxic substances which, like organochlorine pesticides (p. 118) may be accumulated in the tissues particularly of marine mammal like seals, and in marine birds, where they cause serious metabolic upsets including sterility. When this was discovered, industry stopped making PCBs, but considerable dumps remain in many parts of the world. There is no difficulty in destroying PCBs, and transforming them into harmless residues, but creation of suitable plant is expensive. Several firms in Britain have this expertise, which if properly used could soon get rid of all residues. However, when a Canadian firm attempted to import drums of PCBs into Britain, the organisation Greenpeace went to great lengths to stop this, including hampering the ships before they could land, and also by persuading dockers' trade unions to "black" such cargoes. In my opinion this was purely political action, and showed a great lack of understanding of environmental problems. Had the PCBs been landed they would by now have ceased to exist. Refusal to allow the cargoes to land may well result in their being dumped at sea or being improperly treated so that the environment is seriously put at risk.

Nevertheless, at the end of the day there are many waste products of industry which are potential dangers to the environment. These may be relatively non-toxic chemicals that only need to be diluted sufficiently to be harmless, or they may be substances which break down quite quickly in a containment site; but they all require appropriate treatment. Within Britain, our present laws make it obligatory on any manufacturer producing waste to state its exact chemical constitution and what toxic substances it contains. Given this information, there are responsible firms in Britain which can deal successfully with virtually any poisonous substance. The EEC has regulations similar to those operating in Britain, making it obligatory to state the contents of waste for disposal. This information did not exist in the case of the *Karin B*'s cargo. A British firm offered to analyse the drums, and to deal with those substances for which it had facilities. All reputable firms in this field would take the same attitude. Unfortunately there have been instances of small, so-called 'cowboy' firms accepting undescribed toxic waste from abroad and perhaps not dealing with it properly. Provided this is stopped, the importation and destruction of toxic wastes, which is at present widely criticised on environmental grounds, should be encouraged.

The important safeguard of the environment is the site licence, which must be obtained wherever wastes of any kind are disposed of or treated. Responsible contractors, who may be local authorities or commercial companies, adhere very strictly to the conditions of the site licence, otherwise the site may be closed down. Incidentally, these licences are only granted when the operators have the facilities (e.g. analytical laboratories) necessary to show that only approved substances receive the appropriate treatment.

In Britain most of the less dangerous chemical and industrial waste is eliminated by the process known as 'co-disposal', a term which describes the deposition of municipal and certain types of industrial waste in a single landfill site. The procedure has been greatly criticised, and is less commonly used in other European countries. However, the Royal Commission on Environmental Pollution, after careful study of the subject, and after visiting important sites where the operation was being practised, said: 'We conclude that, for many industrial wastes, properly managed co-disposal is an environmentally acceptable option in containment sites.' The same conclusion was reached by parliamentary committees concerned with the environment. Among the various wastes considered unsuitable for such treatment were acid tars, volatile and inflammable organic liquids and drums of toxic substances.

All responsible people involved in managing co-disposal sites will only accept materials of which they know the exact constitution. I myself have been present at such sites when lorries containing unacceptable material, or lorries lacking a proper description of their contents, have been turned away. On one occasion I left the site soon after a load had been rejected, and observed that the driver, unwilling to go the extra miles to a site with the facilities to deal with his load, was tipping it on to the roadside verge. In this case the police were summoned and the culprit was heavily fined as well as being made to remove the material from the roadside.

Co-disposal proceeds as follows. Domestic waste is deposited, compacted, capped with soil and allowed to mature for some time, preferably as long as five years. Then trenches are ripped through the capping material into the body of the now largely decomposed waste. Liquid waste is brought by tanker trucks, and run into the trenches. The liquid should not completely fill the trenches and flood the surface, but even if this happens it is quickly absorbed. After an agreed amount of liquid waste has been absorbed, the capping is restored. Care must be taken to avoid putting liquids which react with one another into the same trench; occasionally the wrong type of mixing has caused the production of poisonous carbon disulphide gas.

The Royal Commission rightly insists that research on co-disposal should continue. It records that when layers of cyanide and phenolic wastes, substances which if liberated could have had seriously damaging effects, were deposited in household waste, only minute amounts were leached out, and that after five years' exposure to rainfall and anaerobic decomposition, more than 95 per cent of the original toxic ingredients had disappeared, mainly by biological conversion into harmless substances.

Critics of co-disposal are concerned that the leachate from such sites will contain large concentrations of heavy metals or other substances contained in the industrial wastes. As these are sites where containment of the leachate occurs, this need not be important. However, many analyses have shown that the municipal refuse, which has acted as a 'sponge' and absorbed the liquid waste, tends to bind metals and many other substances within its mass, so the leachate is generally little different from that produced by simple municipal waste. Nevertheless, all leachate is a potential source of pollution, and care must be taken to see that it does not escape in an untreated state. As already stated, it may be sprayed time and time again on to the surface of the refuse, when evaporation takes place and so the risk of overflowing the banks of the site is reduced. It is also found that when the leachate percolates the refuse for a second time, metals and other substances continue to be absorbed and the leachate becomes weaker. Where large volumes of leachate need to be disposed of, treatment plant is installed, and this will deal with any residues of material co-disposed on the site.

Co-disposal appears to have no harmful effect on the ecology of the waste disposal site, as was indicated when I first compared parts of the Pitsea site, where co-disposal had taken place, with other areas which had only received municipal waste. Two years after the surface had been restored when co-disposal was concluded, the plant cover of the two areas could not be distinguished. Both areas also had similar populations of earthworms or other soil fauna. Furthermore, the extensive lists of plants and animals set out in the appendices to this book were mostly derived from this same co-disposal site; they do not indicate that co-disposal caused any environmental damage. Open water immediately adjacent to the site was examined, and was found to contain a normal and flourishing aquatic fauna. Had pollution outside the site occurred, this freshwater ecosystem would have been the first to react. The conclusion, therefore, is that properly managed co-disposal of appropriate substances does not endanger wildlife.

Some toxic wastes are treated by solidification. The waste liquid is mixed with various lime-containing cements and alumino-silicate compounds called pozzolans to produce a slurry. This sets in a few days, and within six months is rock hard. The process is very much more expensive than co-disposal, and although it is effective it is generally uneconomic, except with dangerous substances unsuitable for deposition along with municipal waste. It is a safe technique, the solidified material may be easily stored (though its volume may be considerable), and there is little risk of toxic materials escaping. But it lacks the great advantage of co-disposal, where most dangerous wastes are not just stored away, but transformed into harmless substances.

It is clear that most toxic waste in Britain today is dealt with successfully, causing little damage to the environment. Most of the concern expressed by the public, and even by parliamentary committees, is therefore unfounded. We have safe methods of disposal, and sites where this can be performed. Damage can only occur if the law is not observed, or if carelessness leads to unnecessary accidents.

I have described some of the old sites which were used before present laws and planning regulations were enforced. The results were horrific, and the fact that there was so little damage outside the sites was largely a matter of luck. Even so, there are instances where the most deplorable practices, such as would not be permitted today, have produced some of the most interesting natural conditions in the country. So, from the conservation viewpoint, improved technology, as in methods of sewage disposal (see p. 103) is not all gain.

Until the beginning of the 20th century, washing soda (sodium carbonate) for the chemical industry and for the housewife was produced from sodium sulphate, coal and limestone by the Leblanc process. This process, widely used throughout Europe, produced serious air pollution such as is almost unknown today in Lancashire (where much of the industry was concentrated), generating many unpleasant and toxic alkaline wastes. These were generally deposited in considerable quantities, sometimes covering an area of some hectares to a depth of several metres. If the dump was adjacent to a stream, this was deprived of all life, assuming there was anything left to kill off as a result of sewage and other industries upstream. On the ground, the wastes, which had a pH approaching 14, were far too alkaline to support any form of plant life, and remained bare and unsightly for many years after they were deposited.

Then gradually, washed by the rain and influenced by the mildly acid carbon dioxide, and possibly affected even more by acid rain containing sulphur dioxide from house chimneys and factories, the free lime (calcium hydroxide) disappeared from the surface layer. This left a very limy soil, containing few nutrients, which allowed calcicole plants to grow with comparatively little competition from other, possibly more aggressive species. These sites have been studied by several botanists, in particular by R.P. Gemmell.

After some 70 years the Leblanc process waste developed the profile described in table 7.

The lime wastes are rich in those species which do well under conditions of high lime content and low fertility. The low nitrogen and phosphorus levels restrict the growth of grasses and clovers which would otherwise suppress the more interesting and uncommon plants. The vegetation is very open or sparse, and allows slow-growing plants, such as orchids, to colonise and form large populations.

Orchids do indeed form the most spectacular feature of these sites. Species which are otherwise only found near the sea, or in other places many miles away, unexpectedly appear, presumably through airborne seeds. Many orchids take up to 15 years from the germination of the seed to flowering, and for most of that period the plants are tiny and obscure, so the appearance of the large flower spike is a pleasant surprise.

The most characteristic species are the Marsh Orchids *Dactylorhiza incarnata, D. majalis* subsp. *praetermissa* and subsp. *purpurella*); but several others occur, including the Common Spotted Orchid (*D. fuchsii*), the Fragrant Orchid (*Gymnadenia conopsea*), the Early Purple Orchid (*Orchis macula*), the Marsh Helleborine (*Epipactis palustris*) and the Pyramidal Orchid (*Anacamptis pyramidalis*). Many orchid hybrids, rare elsewhere, are common. The following hybrids have been identified: *D. fuchsii* × *D. purpurella*, *D. fuchsii* × *D. praetermissa*, *D. fuchsii* × *D. incarnata*, *D. incarnata* × *D. praetermissa*, *D. incarnata* × *D. purpurella* and *D. fuchsii* × *Gymnadenia conopsea*. These hybrids are sometimes much larger than the normal British orchid, in some cases reaching a height of 60 cm.

Although the orchids are the most spectacular, other plants which can tolerate alkaline soil are present. These include some common grasses, for instance Sheep's Fescue (*Festuca ovina*), Red Fescue, Cock's-foot (*Dactylis glomerata*) and Creeping Bent (*Agrostis stolonifera*). As a rule these grow quite well, but are not sufficiently rampant to form a mat and compete seriously with the orchids. Other herbs commonly found are Common Knapweed (*Centaurea nigra*), Wild Angelica (*Angelica sylvestris*), Centaury (*Centaurium erythrea*), the exotic introduced Blue-eyed Grass (*Sisyrinchium bermudiana*), Mouse-ear Hawkweed (*Hieracium pilosella*), Common Hawkweed (*H. vulgatum*), Eyebright (*Euphrasia nemorosa*) and Devil's-bit Scabious (*Succisa pratensis*). The long-term prospect of these sites is somewhat doubtful; they will require suitable management, which mainly consists of ensuring that there is no application of nitrogenous, and particularly of phosphate, fertiliser. Even one application allows grasses and legumes to grow so vigorously that the orchids in particular are seriously diminished. The other risk is that shrubs such as Hawthorn, Goat Willow (*Salix caprea*), Creeping Willow (*S. repens*), Grey Willow (*S. cinerea*) and Silver Birch will appear, and unless controlled may turn the site into dense scrub which, in time, will revert to forest. Indeed, several fairly dense Birch woods already exist.

The Leblanc process ceased to be used in 1919, and was replaced by the Solvay or Ammonia Soda process which produces sodium carbonate from salt, ammonia and limestone. This also generates large amounts of lime waste which has formed some recent lime-rich habitats, but because greater care has been taken in their disposal, these are not as spectacular as those produced in earlier periods from the Leblanc waste.

Although the old Leblanc waste sites are highly alkaline, they may also contain relatively small patches of very acidic soil, as a result of the way in which waste disposal was formerly carried out. Acidic boiler ash was dumped on to the top of the Leblanc waste, and since little mixing took place, the characteristic vegetation of acid soils developed locally. This supports only a few species, typically Wavy Hair Grass

Table 7 Description of the soil profile of the Leblanc process colonised by a lime flora after 70 years' exposure (after Gemmell).

Depth (cm)	pH	Description of 'soil'
0 – 5	7.7	Black with significant humus content.
5 – 15	7.7	Dark with low humus content.
15 – 25	7.8}	Yellowish-brown, markedly stained with ferric deposits.
25 – 35	8.0}	
35 – 45	9.4}	Yellowish, slightly stained with ferric deposits.
45 – 55	9.7}	
55 – 65	12.2}	White with dark lumps of unburnt coal; calcium hydroxide
65 – 75	12.1}	present.
Below 80	12.6	Bluish-white with dark lumps of unburnt coal. Calcium hydroxide present, permanently moist.

(*Deschampsia flexuosa*), Mat-grass (*Nardus stricta*), Common Bent and Sheep's Sorrel (*Rumex acetosella*). The poverty of the fauna, as compared with that on the alkaline areas, is probably explained by the comparatively small areas of acid soil, and the fact that no species such as orchids, with tiny seed easily dispersed by wind, were involved.

Other lime wastes, generally less alkaline than the Leblanc material, arose from lime kilns, limestone tailings and rubble, calcareous sands and limestone ballast. Blast furnace slag, from the smelting of iron ore with limestone to make pig iron, creates a material which weathers to produce something resembling natural limestone. These support typical calcicole plants. However, they are even more liable to be invaded with Hawthorn and Willow than are the Leblanc wastes, and the improved fertility caused by leaf fall and humus enrichment encourages further scrub and grass, and gradually eliminates the pioneer calcicole plants.

I have already described (p. 51) how power station ash (PFA) is treated to produce grass fields and even agricultural land. In its raw state, the PFA generally contains toxic levels of boron and other metals. Nevertheless, ash with a higher than usual lime content is occasionally produced, and here interesting plants may appear. At one ash site in north-west England dense populations of several orchids, including Marsh Orchids, Common Spotted Orchids and Marsh Helleborines, appeared within a period of 15 years from the original tipping. As willow scrub also flourished, this site, unless skilfully managed, will quite soon lose its interest.

Inland salt marshes have occasionally developed around natural brine springs, but the salt industry, as a result of brine spillage and subsidence caused by underground pumping of brine, has brought about the local development of brine pools and salt marshes. As these are some distance inland, the seed of most salt-tolerant species (halophytes) has not reached them. Species which have been found in these situations include Orache (*Atriplex patula*), Salt Marsh-grass (*Puccinella distans*), Sea Spurrey (*Spergularia marina*), Sea Aster (*Aster tripolium*), Sea Arrowgrass (*Triglochin maritima*) and Sea Club-rush (*Scirpus maritimus*). Other areas rich in salt are roadside verges where the carriageway has been heavily treated with de-icing materials. There is evidence of halophytic grasses colonising these verges, starting at the seaside and progressing inland. It is thought that moving vehicles transport the seed.

As described in Chapter 4, colliery spoil is not well colonised, particularly where it is very acidic, because of the content of pyrites, which release sulphuric acid over periods of more than 100 years. In some cases the only obvious coloniser is a poor stand of Wavy Hair Grass (*Deschampsia flexuosa*). Less acidic heaps may grow Mat-grass, Common Bent, Yorkshire Fog (*Holcus lanatus*) and Creeping Soft-grass. In time some herbs, including Sheep's Sorrel, appear. Eventually willows and birches may establish themselves, to be followed by Sessile Oak (*Quercus petraea*) and Aspen (*Populus pendula*), with Bramble (*Rubus fruticosus*) both under the trees and in open ground. This shows that, given time, even the most inhospitable areas develop some interesting vegetation. It is perhaps a pity that today, when we are so much more careful not to leave large quantities of toxic wastes lying about the countryside, we will never again produce anything comparable to the orchid-rich areas that resulted from the almost criminally careless dumping of Leblanc waste.

Chapter 6

AIR POLLUTION

The atmosphere of the troposphere, the layer of air supporting all plant and animal life and covering the globe to a height of approximately 10 km, consists of 21 per cent oxygen, 78 per cent nitrogen and 0.033 per cent carbon dioxide. It also includes about 1 per cent of argon, neon, helium and other inert gases which, as far as living organisms are concerned, act in much the same way as nitrogen. There is a further gas which, chemically, falls into this class, but which is possibly more important. This is radon, present in very small amounts (parts per thousand million) which would not have been detected were the gas not radioactive. This gas is produced naturally out of granitic rocks, particularly in Cornwall. It is not, strictly speaking, a waste material, but is considered in Chapter 10 along with radioactive waste products which may have environmental effects.

Pollution seldom affects the basic constitution of the air, say by reducing the oxygen available for respiration. A significant reduction may occur in a confined space like a mine, but not in a crowded room even with the windows closed. People who complain of lack of oxygen are actually reacting to heat, to the smell of the other occupants, and, in extreme cases, to a rise in the level of carbon dioxide. Altitude has a greater effect. On the top of Ben Nevis the oxygen level is reduced by about a sixth, a fall that is hardly detected by a healthy individual. At greater heights, as in unpressurised aeroplanes, oxygen deficiency can be more serious. In the Andes at over 4,000 metres above sea level, many unacclimatised visitors are seriously distressed, and extra oxygen has been used by most successful climbers of Everest. Pollutants are usually present in very small amounts – parts per million, per thousand million or even less. The important thing is that they have extremely significant effects at these low levels.

When man first made fire, the smoke (the waste product of the combustion of wood) polluted the atmosphere. While human numbers were small, and life primitive, little smoke was produced, and it did not do much damage to the environment, to man himself, or to other organisms.

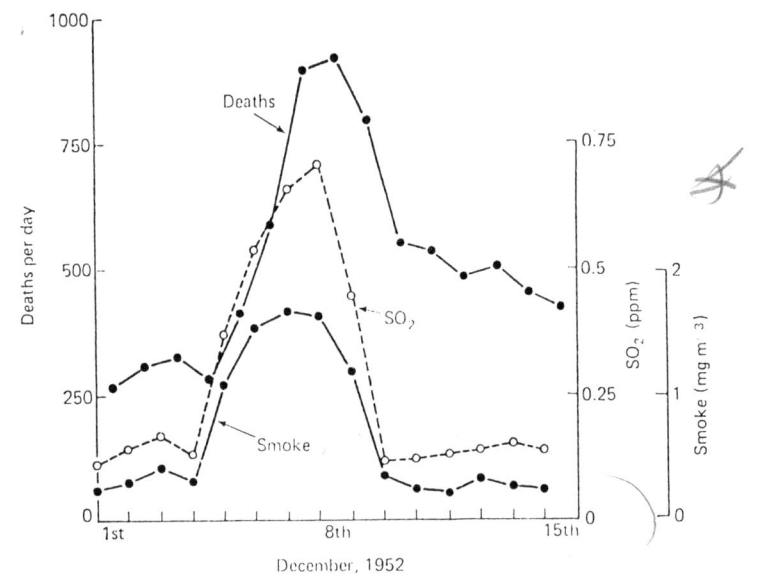

Fig. 5 Excess deaths and pollution levels in the great smog of 5-9 December 1952 (from Royal College of Physicians 1970).

However, by the 19th century, smoke was the most serious cause of urban pollution. London 'pea soup' fogs are described by many Victorian authors. Charles Dickens, in *Bleak House*, gives a graphic description of a 'London particular'. The situation was already beginning to improve when, in 1952, the true danger of this type of pollution was finally realised. Throughout every winter London suffered from pollution by smoke, but this was relatively bearable unless climatic conditions exacerbated the situation. Between 5 and 9 December 1952 there was a period of calm weather with a high barometric pressure and a temperature inversion which trapped the smoke and other pollutants near to the ground. The mixture of smoke and fog became known as 'smog'. This had often happened before, when conditions must on occasion have been even worse, but no one had been particularly worried, because there were no statistics to indicate what was happening to the population of London. In 1952 figures for deaths were available from the Registrar General's Office, and these showed that there was an excess of 2,484 deaths during the four days of the Great Smog (Fig. 5). Those who died were mostly elderly individuals suffering from bronchial complaints, many of whom might not have survived many months longer even in smogless conditions but, nevertheless, these deaths shocked the country.

The government was quick to act. It immediately set up a

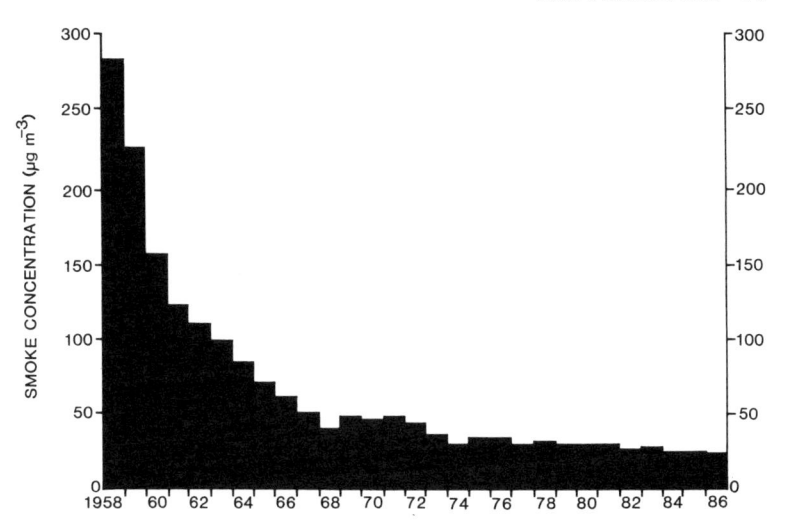

Fig. 6 Average smoke concentrations near ground level in London 1958-86 (Data from Warren Spring Laboratory)

Commission to investigate the situation, and in 1956 passed the Clean Air Act. This controlled the output of smoke from industry. More important, it led local authorities to set up 'smokeless zones' in cities and towns, in which the burning of raw coal was forbidden. The result was dramatic, as indicated in Fig. 6 which shows the ground level smoke concentrations in London following the passing of the Clean Air Act. Similar figures could be produced for most of our cities.

The result of the fall in smoke levels on human health was remarkable. During the 1960s there were several episodes when temperature inversions similar to that of 1952 took place. There was no noticeable smog, and no significant increase in deaths. Fig. 6 shows that levels of sulphur dioxide rose in parallel with the smoke, but to an even greater extent. As will be seen (p. 73), while smoke levels fell after 1956, sulphur dioxide levels continued to rise, and were higher in London during the 1960s when these periods of inversions occurred. So humans can evidently tolerate quite high levels of sulphur dioxide alone. The deaths in 1952 were probably caused by a combination of pollutants, smoke, sulphur dioxide, and possibly sulphuric acid which was also present in rather small amounts.

The smog deaths in cities were to some extent self-inflicted. Humans were both victims and perpetrators. There is little evidence that wildlife suffered, though there are reports that some of the prize cattle at the Smithfield show in London also died in 1952.

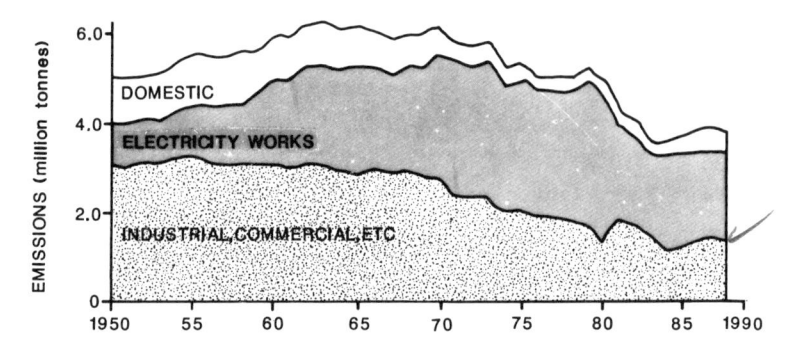

Fig. 7 SO$_2$ emissions in the UK 1960-86. (Data from the Warren Spring Laboratory)

Smoke was primarily an urban problem, either being deposited within the built-up area, or diluted substantially when it reached the open country, so that the effects on rural areas and on wildlife in general were limited. In the Peak District, between Manchester and Sheffield, there was sufficient smoke deposited to blacken the sheep and the clothing of people sitting on the heather. Trees and walls were also blackened in most urban and suburban districts. Perhaps the most notable effect of the smoke was to encourage the selection of melanic forms of certain insects. The best known is the Peppered Moth (*Biston betularia*) which has been described in detail by E.B. Ford in his book *Ecological Genetics*. This moth has two races, the typical pale form and the much darker melanic race. In clean areas, with little pollution from smoke, the tree trunks are covered with pale lichens, and the typical form of moth is well camouflaged. The melanic form is very prominent. This difference is particularly obvious to birds which prey on the insects. In industrial areas, where the tree trunks are blackened and lichens are rare, the melanic form is inconspicuous and the typical form more visible. The typical form is thus restricted mainly to clean rural surroundings, and the melanic to the more polluted urban districts. Since smoke levels in towns have fallen, the typical form has reappeared there, and the melanic race has become less common. Melanism in towns has been found in other insects, too, particularly ladybird beetles (*Coccinellidae*).

Probably the most serious waste product which gives rise to air pollution is sulphur, mostly in the form of sulphur dioxide. This gas is deposited directly on plants, buildings and the soil ('dry deposition') and, at least near the source of emission, only a small fraction is dissolved in the rain. As mentioned above, output from all sources in Britain

AIR POLLUTION 73

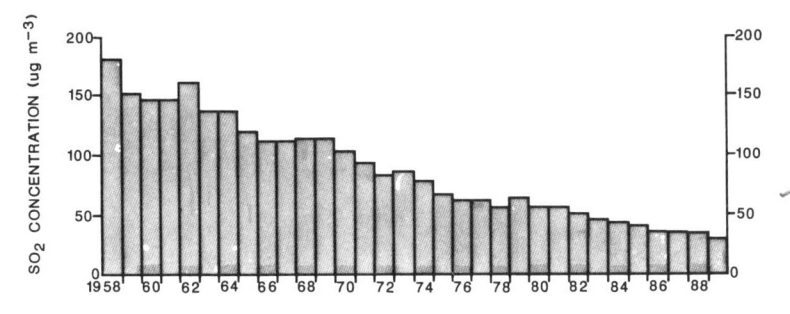

Fig. 8 Average SO_2 concentrations near ground level in the UK 1958-86. (Data from Warren Spring Laboratory)

continued to rise for some 15 years after the passage of the Clean Air Act 1956, exceeding 6 million tonnes in the early 1960s. After that the output fell to 3.58 million tonnes in 1985. As Fig. 7 shows, the reduction was caused by a great decrease from domestic sources, as a result of coal being replaced by natural gas (containing little sulphur) and electricity, whose sulphur output was found elsewhere. Industry also converted largely to fuels which did not produce sulphur dioxide. By 1985 75 per cent of the sulphur dioxide came from the coal-fired power stations of the Central Electricity Generating Board. British coal, on average, contains some 3 per cent sulphur, and this causes virtually all the output to the atmosphere.

Fig. 8 shows the average ground level concentrations of sulphur dioxide for the whole of Britain. It records the substantial fall in this level, but does not show the great variations found in different locations. In London the average sulphur dioxide concentration in the 1950s was about 200 micrograms per cubic metre (0.06 parts per million), but on occasions, when there was a temperature inversion, this could rise for a short time to 3,000 micrograms or one milligram per cubic metre (1 part per million). In 1985 the average level for London was 50 micrograms per cubic metre, though short time surges to maximum levels as high as 1,000 were sometimes observed.

High levels of sulphur dioxide had considerable effects on living organisms and inert objects. Silver in cities blackened in a few days, and curtains near open windows rotted in a matter of weeks. Humans appeared to be comparatively resistant – experiments in which volunteers were exposed to sulphur dioxide levels of 3,000 micrograms per cubic metre showed few harmful effects. Plants reacted more strongly. Until recently it was impossible to grow many species of coniferous trees in city parks and gardens. Fungi were even more seriously affected. This

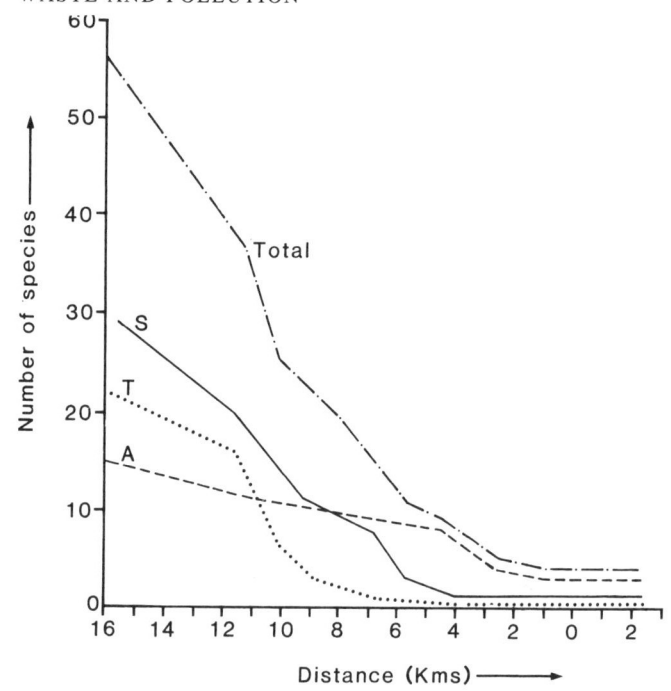

Fig. 9 Number of species of lichens occurring on asbestos- cement (A), sandstone (S) and trees (T) along a transect through the centre of Newcastle upon Tyne (After Gilbert)

was not always considered a disadvantage. Roses grown in cities were not themselves harmed by the sulphur but neither did they suffer from blackspot fungus disease, *Actinonema rosae*, which was much more susceptible. Today (1988), with the falling levels of pollution, the disease has reappeared. The tar spot disease *Rhytisma acerinim* which attacks sycamore trees, was also found to be absent from locations with high sulphur dioxide concentrations. The relationship of the disease to sulphur dioxide concentrations, as determined by R.J. Bevan and G.N. Greenhalgh, is shown in Table 8.

Lichens have been found to be particularly susceptible to damage from sulphur dioxide and the diverse species are damaged by different concentrations of the gas. This has enabled lichens to be used to map the levels of sulphur dioxide pollution throughout the country. Lichens are composed of two different organisms, i.e. a fungus living in symbiosis with an alga. They have no proper roots through which water is taken up, depending for their supplies of water and salts on what is contained in rainwater, which either falls on them directly or on the substrate to

AIR POLLUTION 75

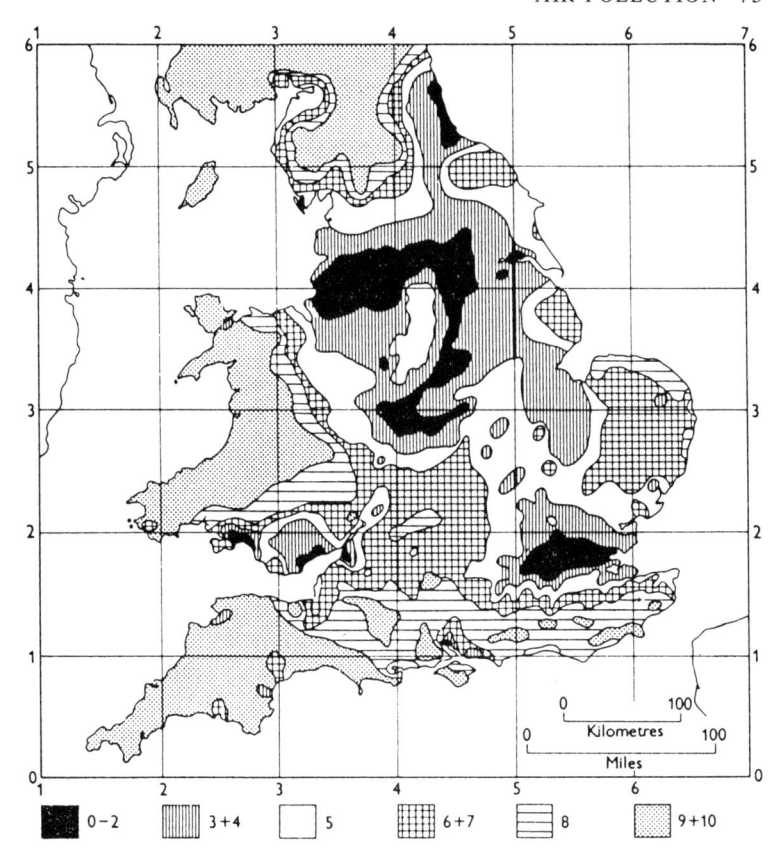

Fig. 10 Approximate limits of lichen zones in England and Wales. Further details in Table 9 (p. 76). (After Hawksworth and Rose)

which they are attached. The distribution of lichens is thus governed by the levels of sulphur dioxide in the air, and by the nature of the substrate. Lichens do better on basic or nutrient-enriched bark (e.g. elm) than on moderately acid bark (e.g. oak). Sandstone, being comparatively alkaline, may support the largest number of species.

Pollution by sulphur dioxide was generally greatest in the centre of a city, and because levels fell as one moved out to the country, so there was a gradient of lichens. Fig. 9 shows what was found by O.L. Gilbert in Newcastle-upon-Tyne. One interesting fact is that in the more highly polluted zone up to some 8 km from the centre, the largest number of species was found on the asbestos-cement, but in the purer air further out, this total was rapidly exceeded by the population of the sandstone and, even further out, on the trees.

76 WASTE AND POLLUTION

Table 8 Tar spot and sulphur dioxide levels.

Severity of disease	Annual average SO_2 ($\mu g\ m^{-3}$)
No tar spot, leaves clean	Above 85
Tar spots less than one per 100 cm leaf area	75–85
Average 10 spots per tree	55–70
Tar spots frequent, 2 or more per leaf common	40–55
Most leaves with 1 to 4 spots	23–40
Tar spots on all levels, many spots per leaf	Less than 25

Table 9 Lichen zone scale, devised by O.L. Gilbert

Zone	Lichens and mosses present	Mean winter SO_2 ($\mu g\ m^{-3}$)
0	Pleurococcus on sycamore	Over 170
1	Lecanora conizaeoides on trees and acid stone	About 150-160
2	Xanthoria parietina appears on concrete, asbestos and limestone	About 125
3	Parmelia (Saxatilis sulcata)appears on acid stone and Grimmia pulvinata on limestone or near mortar	About 100
4	Grey leafy (foliose) species (e.g. Hypogymnia physodes) starts to appear on trees	About 70
5	Shrubby (fruticose) lichens (e.g. Evernia prunastri) appears on trees	About 40-60
6	Usnea becomes abundant	35 or below

Gilbert first devised a scale, based on the lichens present, to set up seven zones throughout the country depending on the lichens present. This proved sufficiently easy to use so that it was developed as an exercise for school children studying air pollution (Table 9). D.L. Hawksworth and F. Rose further refined the system, and extended it to ten zones. These are depicted in the map (Fig. 10) and in the extensive Table 11). For those not sufficiently expert to determine the exact species, some approximation can be made by noting that the further the lichen protrudes from the substrate, the more susceptible it is to sulphur dioxide damage.

The British Lichen Society has produced an atlas of lichen distribution, which shows how some species declined with increasing pollution in the period up to 1960, and how others, more tolerant of sulphur dioxide, increased. More recently, M.R.D. Seward has found that in some areas near towns where sulphur dioxide levels have fallen substantially, some of the more delicate species are beginning to return. The process is slow, growth may take a long time, but it does suggest that we are gradually returning to the lichen distribution which existed 200 years ago.

There are many other records of damage to plants from the high levels of sulphur dioxide which existed until comparatively recently. The Pennine range received the emissions from industrial Lancashire, and these were sufficient to make it impossible to grow coniferous trees. Today's low pollution levels now make it possible. This shows that the soil has not been irreversibly damaged by the earlier high levels of pollution. There are many similar instances of the lethal effect of sulphur dioxide in forests in Czechoslovakia, Poland and eastern parts of Germany. Some grasses have also been found to be damaged in parts of Britain. On the other hand, there are soils which are deficient in sulphur, which is an element essential for life, and in some such cases the 'pollution' has had a beneficial effect.

As sulphur dioxide, generated in towns and industrial areas, spread over the countryside, it mixed with the air and became considerably diluted. Until recently it was generally believed that when diluted to levels such as 25 micrograms per cubic metre, it would cease to have any harmful effect. For this reason the accepted method of disposal was 'dilute and disperse', which usually proved satisfactory. One remarkably successful technique in reducing levels in towns and near to sources of emission was to build chimneys up to 200 m high, in many situations where local sulphur dioxide damage had previously been serious. The flue gases left the chimney at a considerable velocity, and quickly mixed with air well above the top of the chimney.

Complaints, including those from the EEC (who should know better), that the use of high chimneys is simply transferring pollution from the place where it is produced to distant sites which are thus made to suffer, are not strictly justified. At 100 miles or more from the source, sulphur dioxide levels are much the same whatever the height of the chimney. This is because the amount of gas which builds up dangerously high concentrations near the source is only a small percentage of the total. Even with low chimneys, and severe local damage, the greater part of the sulphur dioxide still finds its way into the atmosphere and is diluted and blown away in much the same amounts as the gas emerging from the tall chimneys.

Unfortunately, sulphur dioxide and other air pollutants are not always dispersed in the manner just described. Sometimes great 'slugs' of pollutants remain together at quite high concentrations, and may cause unexpected damage some way from source. For this reason there is growing belief that the 'dilute and disperse' technique is not always reliable, and that more efforts should be made to reduce emissions at source.

Nor have high chimneys always been successful in preventing local pollution. At Glossop in the Peak District there is a factory producing metal alloys which emits considerable amounts of sulphur dioxide. With the intention of stopping local pollution, a chimney some 100 m high has been erected. The local people complain that this has made things worse – the pollution is blamed on 'the chimney'. It is probable that at most times the chimney does get rid of much of the pollution, but not always. Glossop is in a deep valley surrounded by hills which are higher than the top of the chimney; and when the wind is in the wrong direction it blows the plume down on to the town. This indicates that no single method works in all situations. The only satisfactory solution here would be to trap the pollution before it entered the chimney.

We have discovered in recent years that the very dilute sulphur pollution, blown hundreds or even thousands of miles from its source, may still have important effects on freshwater and living organisms. This is because of what is commonly spoken of as 'acid rain'. As will be seen, this term is used to cover a whole series of different reactions which are related to different forms of air pollution, and it has given rise to a complicated web of fact and fiction, causing a great deal of confusion.

The central reaction is based on the fate of the sulphur dioxide, which at high concentrations is so very phytotoxic, and which when sufficiently dilute is either harmless or positively beneficial. In the atmosphere, complicated chemical reactions take place, some quickly, some slowly. The result is that, perhaps over a period of several days, during which the air travels great distances, sulphur dioxide is transformed to sulphate, which may finally yield sulphuric acid. Because sulphate is less toxic than sulphur dioxide, this reaction was generally accepted as nature's method of detoxification; and under certain circumstances, this does happen. But frequently the sulphuric acid, so dilute that it causes no damage even to susceptible plants, is washed out by rain and penetrates the soil. In places where the rainfall is very heavy, a considerable amount of acidity builds up, and where the soil has a poor buffering capacity this may have quite serious effects.

The sulphuric acid produced in this manner produces genuinely acid rain – but not, as a rule, very acid rain. The phenomenon of acid rain, in fact, is not a new one. Actually, all rain, unless immediately adjacent to a site such as a chalk quarry where there is alkaline dust, or a great concentration of cattle emitting ammonia, is acidic, as the carbon dioxide itself produces a pH of 5.6, and other naturally occurring substances often reduce the pH to somewhere in the region of 5. However, man-made acid rain was recognised by R.A. Smith, the First Alkali inspector, who did so much to curb the output of hydrochloric acid from alkali works. In 1852 he published his classic paper 'On the air and rain of Manchester', which showed that many pollutants – hydrochloric acid as well as sulphur compounds – produced rain with a lower pH than that complained of today. However, Smith's acid rain was caused by primary pollutants close to where they were being emitted. The acid rain that causes concern today contains secondary pollutants, substances differing from the primary pollutants which human activities produced in the first instance.

One phenomenon regarding acid rain commands general agreement. Where rainfall is very heavy, even if it is not strongly acidic, a considerable amount of acid ions enters the soil. If such soil is derived from granitic rocks, and contains little calcium, sulphur eventually builds up in it. Streams and lakes fed from the water that percolates through the soil become acidic and may no longer be able to support fish, particularly species like Trout and Salmon which are most seriously affected by acid conditions. This situation, in which air pollution leads to changes of an unfavourable nature in fresh water, is discussed in Chapter 7.

The greatest concern about the harmful effects of so-called acid rain is to trees, particularly in Germany where the possible danger of *Waldsterben* ('forest death') has received much publicity, particularly from 'Green' politicians. Some of our so-called environmentalists in Britain consider that the same phenomenon is rife here. The difficulty is to discover exactly what is happening to our trees, and to those which are being studied in Germany and Scandinavia. Only when this is done can we begin to investigate the cause of any damage which actually occurs. The situation is complicated because some people seem positively to enjoy reporting damage and destruction, even when a more careful long-term study of the trees indicates that their reports have been greatly exaggerated.

There is no doubt, however, that many trees have been seriously damaged, and some have died, in Germany and in Central Europe, in recent years. It is equally evident that there are sick and dying trees of several species in Britain. The view most often expressed by the media is that this is all our fault, in particular of the Central Electricity Generating Board, which produces vast amounts of poisonous sulphur dioxide. This is then transformed into sulphuric acid which, it is alleged, is mainly deposited abroad where it causes much death and destruction. The trouble about this sort of statement is that it contains some elements of truth, but gives a very false overall picture of what is happening to forests in many countries.

Damage to trees, particularly conifers, by sulphur dioxide has been well known for a long time. We have seen that until recently it prevented trees from being grown in the Pennine region of Britain. Where sulphur dioxide, as a primary pollutant, occurs in sufficient concentration, trees die. This is happening today, and the source of the gas is mainly Czechoslovakia, Poland and the USSR. There is very little damage from this cause in Britain today. One easy way of detecting high concentrations of sulphur dioxide, which has been described above (p. 74) is by the presence and absence of lichens, in particular shrubby species. In some parts of Europe, where sulphur dioxide is indeed the culprit, such lichens are almost totally absent. In the German Black Forest lichens are particularly abundant, as they are in most parts of Scandinavia. So, if damage occurs, some other cause must be found.

One difficulty is to decide just what should be termed 'damage'. This problem has been studied particularly in Germany with Norway Spruce (*Picea abies*), which makes up nearly 40 per cent of the forests, and is the species with the greatest economic importance. Before 1984, in Germany, Spruce were considered as damaged if they had lost 10 per cent of their needles; using this criterion, one-third of the trees were described as damaged. Further work showed that there was considerable variability in trees which were otherwise quite normal and many quite healthy trees may drop over 10 per cent of their needles. Today any foliar loss of over 10 and under 25 per cent is considered to indicate an early warning of possible later damage (Fruhwarnstufe). In the early surveys, the authorities asked those making the examinations to include only damage caused by air pollution. This gave some worthless results, as no one, and certainly not the foresters involved, knew exactly what was causing the damage, but the results received wide publicity as proof of the severe damage from air pollution or 'acid rain'. The new classification suggests that less than 20 per cent of the forest area has suffered damage, as contrasted with over 50 per cent, using the old method. Some critics suggest that this change has been made to conceal the real extent of the catastrophe which has struck the forests, but most competent scientists accept the new classification as accurate.

The great worry, in the early 1980s, was that there was rapidly increasing damage to both conifers and broad-leaved trees in Germany – hence the talk of *neuartige Waldschaden.* Fortunately the amount of damage has increased little if at all from 1984 to 1987. What is most encouraging is that although damage is still sufficient to worry everyone concerned, there have been significant signs of recovery, in the Black Forest and parts of lowland Bavaria, by Fir, Spruce and Pine. Moreover, economic damage has not been as serious as was feared. Damaged and healthy trees have the same physical characteristics, and command the same price in the timber market. Tree ring studies show that growth depends predominantly on climatic factors, and that until more than 30 per cent of the foliage is lost, growth continues normally.

Surveys in Britain indicate that foliar loss, as in Germany, is not uncommon, but nevertheless the trees are probably reasonably 'normal'.

There is no evidence that air pollution is having a serious harmful effect. There was some concern about Beech trees. In 1985 a survey, carried out and much publicised by Friends of the Earth in early autumn, showed that many were developing autumn colour and losing their leaves several weeks earlier than usual. Some people claimed that this was caused by pollution, though the probable cause was the very dry summer – periods of drought usually affect shallow-rooted Beech. In 1986 and 1987 the early leaf loss was not repeated; in fact, near Cambridge the large Beech trees on the Gog Magog hills could be picked out from a distance in early October by their dark green coloration. Then when the gale of 16 October 1987 struck, Beeches were more badly affected than other trees, largely because they offered such great wind resistance as their leaves were still stuck on so tightly to the branches – thus they suffered because they were so healthy!

Nevertheless, it would be unwise to minimise the damage which has taken place on the continent and efforts must continue to find out its cause or, more likely, causes. Historical records show that serious damage has occurred on a number of past occasions, but it has not been so widespread nor has it happened at approximately the same time in such widely dispersed locations. It must be conceded that all the causes of all the different types of damage have not as yet been identified. There is little doubt that there have been several different pathological symptoms, and that these have certainly not all had a common origin.

Taking Norway Spruce as an example, five different types of damage (or 'decline', to use the current jargon) have been identified. These are needle yellowing at higher elevations, thinning of tree crowns at medium-high altitudes, needle reddening of older stands in southern Germany, chlorosis of needles in the Bavarian Alps, and thinning of tree crowns in northern coastal areas. Various causes of these symptoms have been suggested. Multiple stress has been related to damage in general, as has soil acidification and aluminium toxicity. Magnesium deficiency would seem important where needle yellowing is prevalent, and excess nitrogen deposition is suggested as the cause of crown thinning in northern coastal areas, and possibly a factor in other types of damage. It is impossible to test multiple stress, with contributions from air pollution, nutrient deposition, growth alteration, toxic substances, and an array of climatic stress and contributing pathogens; and although this may be the cause of damage under some circumstances, research is tending to concern other factors. The hypothesis that ozone, acting with acid mist, and probably with climatic factors, was at one time popular, but observations in the field do not correspond well with those produced in experimental chambers. Magnesium deficiency, combined with a series of dry years, is apparently the most likely cause of some damage, except in countries such as Britain and Norway where there is considerable magnesium deposition from maritime sources.

The hypothesis that excess nitrogen deposition greater than the trees' requirements may increase susceptibility to climatic and biotic stress, and accelerate soil leaching, is receiving increased support. These effects are said to occur in pines in the Netherlands where there is excessive nitrogen deposition mainly in the form of ammonia from animal waste. This may cause thinning of tree crowns in these northern coastal areas. In central and southern Germany lower, though still elevated, levels of nitrogen deposition may increase tree growth, sometimes causing magnesium deficiency when supplies in the soil are inadequate to meet these increased requirements.

It is clear that we cannot give short, snappy answers regarding all types of damage to trees. However, recent advances in research lead to the following generally accepted conclusions:

1. A more realistic definition of damage takes the natural variability of tree condition into account and sets a generally accepted threshold for damage of a foliage loss over 25 per cent.
2. Initial projections of continuously increasing damage have not been borne out. There is, on the contrary, regional recovery.
3. It is generally accepted that new-type forest damage is not a single phenomenon, but a series of regional damage types, possibly triggered by a common synchronising factor, so far not certainly identified.
4. The causes of some individual damage types are partly resolved, as detailed above.

When we come to examine the British situation, certain points seem to be clear. We may have local damage, and isolated trees may be suffering, but we have nothing corresponding to the worst patches of new-type forest damage that are found locally in parts of Germany. The one thing which is not causing widespread damage is true 'acid rain'. There is no reason to fear that any substantial number of our trees are in any way 'doomed' and, with suitable management, our forest cover should flourish for many years to come.

Air pollution by sulphur has had one other important environmental effect. It has been responsible for the erosion of buildings and stone, particularly limestone where the calcium carbonate is turned into sulphate. Over geological time, the main cause of the erosion of stone has been the carbonic acid in the rain, arising from the carbon dioxide universally present. Stone near the sea, where salt is present in the air, also erodes more rapidly than inland. There is little doubt that the high urban levels of sulphur dioxide present in cities has affected stonework in buildings

AIR POLLUTION   83

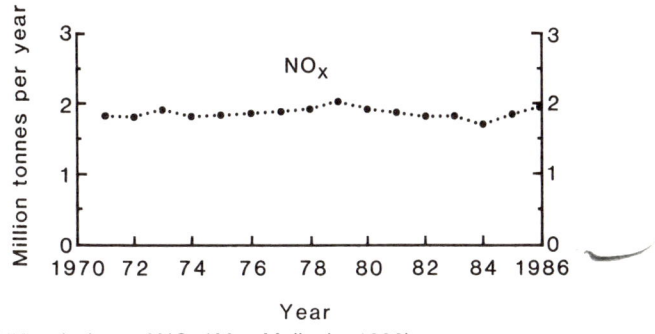

Fig. 11   UK emissions of NO$_x$ (After Mellanby 1988)

like Westminster Abbey, where as long as 100 years ago much of the exterior carved stone had to be replaced. It was generally expected that as soon as sulphur dioxide levels in cities fell (as they have fallen in the last 25 years) damage to cathedrals and other buildings would cease to be important. Unfortunately this has not happened, but it is likely that much of the damage noted today is due to the 'memory effect'. Stone was probably impregnated with sulphur dioxide which slowly rotted it away, but the damage was not visible until recently when the weakened surface began to break up. As far as we can judge, it is the dry deposition of sulphur dioxide which affects the stonework, and wet deposition (i.e. acid rain) as such does little damage.

After sulphur dioxide, the next most important emission of waste gases is that of the oxides of nitrogen. We tend to speak of these as NOx, as it is impossible to be sure of the exact constitution of the gases at any one moment. Chemical reactions are always going on, and as these are comparatively rapid (much quicker than the transformation of sulphur dioxide to sulphate) it seems easiest to use this notation. As Fig. 11 shows, in 1985 total emissions of nitrogen oxides in Britain amounted to 1.84 million tonnes. 40 per cent of this came from electric power stations, a further 40 per cent from road transport, and the remaining fifth of the total from other industries, domestic and other sources.

When air, consisting mainly of nitrogen (78 per cent) and oxygen (21 per cent) is heated to above 1100°C, oxides of nitrogen, particularly nitric oxide (NO) are produced. This happens in car engines, particularly at high speeds. It also occurs in power stations. NO is the main oxide of nitrogen produced under these circumstances. Levels exceeding 0.1 ppm have been measured in Los Angeles, California, but usually levels are substantially lower. At these concentrations NO appears to be harmless to humans, to other animals and to vegetation. The trouble is that in bright sunlight, NO is turned into nitrogen dioxide, NO$_2$. This is much more toxic. It is a brown, evil-smelling gas which is very phytotoxic, and which also causes human respiratory upsets. It is the cause of the brown smog seen in Los Angeles and many cities throughout the world. Nitrogen dioxide levels in Britain seldom seem to reach danger levels, probably because we have insufficient sunshine. The production of nitrogen dioxide by the following reaction in which ozone plays a part:

$$NO + O_3 = NO_2 + O_2$$

A whole series of other reactions takes place, which have as their end point nitric acid, which contributes substantially to the acidity of the rain.

We are still unable to quantify the exact effects of air and rain pollution by oxides of nitrogen and nitric acid. In parts of Europe, particularly Switzerland, it is thought to be the major cause of forest decline. It is important in its contribution to the additional nitrate, the effects of which are discussed above. This may be entirely beneficial, promoting growth, or it may increase the susceptibility of the trees to climatic and biotic stress. The nitrate may be absorbed by the soil and make less contribution to freshwater acidity than might have been expected. Nevertheless, it is probably wise for industry to try to limit its output, and for motor vehicles also to reduce their effect. Current attempts in Germany to lower speed limits to this end are unlikely to succeed as German drivers on autobahns behind the wheels of large Mercedes are very reluctant to remove their feet from the accelerator pedal. The faster a car is driven, the more NOx it produces.

Nitrogen dioxide is potentially dangerous, but it also helps to produce, and for that matter destroy, ozone, the importance of which is becoming increasingly evident. In sunlight, ozone is produced; overnight it is destroyed and nitric acid is produced. Under the climatic conditions obtaining in Britain, vehicles and power stations, on balance, probably reduce the amount of ozone in the atmosphere, but in sunnier climes nitrogen dioxide makes a major contribution. High levels of ozone, over 0.2 ppm, are phytotoxic, and cause serious crop damage, particularly in America. In Britain leaf damage has been detected in very sunny weather on sensitive varieties of tobacco, but not in normal farm crops or wild vegetation. However, if our climate changes, as is suggested by those studying the so-called greenhouse effect, then ozone levels may become higher and may cause more serious damage.

I have not, in this chapter, dealt with one important topic concerning the effects of air pollution. We know that the effect of two different pollutants, for instance sulphur dioxide and nitrogen dioxide, acting simultaneously, may be complicated. Sometimes the effect is simply additive, the damage from one pollutant adding to that of the other. The effect may be synergistic, in that the combined effect may be much greater than additive. And we sometimes find an antagonistic effect, one pollutant actually reducing the effect of the other. My reason for omitting these matters is that we know too little about the effects on the wild vegetation of Britain. Nevertheless, I have no doubt that unless we succeed in greatly reducing pollution levels (and this is possible, for we have moved quite a long way in that direction), further work on pollution mixtures may change our views on environmental management.

There has been much concern about the polluting effects of lead in petrol. It is true that lead is a serious problem, and with soft water dangerous levels can be produced in houses with lead pipes. Lead in paint, particularly when it is old and crumbling, can also damage children suffering from 'pica' (defined as 'eating of substances other than normal food') who consume dust and who chew painted wood. Although there is little evidence that children have been harmed by the lead in petrol, cars can run almost as well without it, and it is therefore being removed as a precaution. Nothing indicates, either, that any form of wildlife has been harmed by the lead from cars. A narrow strip of grass, perhaps 2 m wide, on the side of busy roads shows raised lead levels in the soil and adhering to the grass. Voles trapped there do show some elevation in body levels of lead, but not sufficient to have any adverse effect. Animals trapped further away seem quite unaffected.

The waste gas produced in the largest amounts is actually carbon monoxide. In 1985 5.39 million tonnes were voided into the atmosphere, 84 per cent of this coming from road transport, almost entirely from petrol-burning motors, as diesel engines produce very little. Carbon monoxide is, potentially, very dangerous. It combines with haemoglobin, the red pigment found in our blood and that of all mammals and birds. Quite low concentrations of carbon monoxide can interfere with the transport by our blood of oxygen to our tissues. Many deaths from carbon monoxide poisoning occur every year, often because this is used as a method of committing suicide. However, except in enclosed spaces, carbon monoxide seldom seems to have any serious effect. Policemen on point duty in heavy traffic have not been found to have more than 4 per cent of their haemoglobin put out of action, and this level does not seem to affect behaviour or efficiency. Much higher levels of carboxyhaemoglobin (the substance produced when carbon monoxide is breathed) are found in cigarette smokers, without any obvious effects.

Chapter 7

WATER POLLUTION

All animals require access to water, and man is no exception. We go to considerable lengths to provide a clean and unpolluted supply of drinking water. We also indulge in many activities which pollute the limited amount of fresh water in our country. Water is endangered in a number of different ways, which I shall briefly summarise.

When organic matter is deposited in water, and this may be the natural dropping of leaves from trees or the discharge of excrement from human settlements, the decomposition of the organic matter and the consequent chemical reactions may completely deoxygenate the water, and make it impossible for any life, excepting some anaerobic bacteria, to continue. At best, water contains only a rather low level of oxygen. Thus at 20°C a litre of water saturated with oxygen only contains some 6 cc of that gas, while a litre of air contains 210 cc. The warmer the water, the lower the volume of oxygen it can dissolve, as is discussed in Chapter 8. This reduction in oxygen levels at higher temperatures presents fish and other aquatic life with a problem, for at higher temperatures cold-blooded animals have a greater rate of metabolism and therefore need more, and not less, oxygen. Incidentally, when an air-breathing animal needs to take up and use a litre of oxygen, it has to breathe in only some 5 litres of air. An aquatic animal, which 'breathes' water (a fish, for example, passes the water through its gills), in order to obtain one litre of oxygen would need about 200 litres or very much more if, as often occurs, the water is only partially saturated. This makes aquatic animals particularly susceptible to poisons dissolved in the water, as they make intimate contact with a volume which may contain an appreciable amount of a toxic substance present only at a considerable dilution. Also, water containing chemical pollutants is likely to be partially deoxygenated, so the problem gets even worse.

As already mentioned, toxic substances, wastes from many human activities, may be discharged into fresh water and taken up by animals living in the water. They may be absorbed by water plants, and so affect herbivores. They may, where carnivores are concerned, also pass through quite complicated food chains or food webs, perhaps becoming concentrated in levels much higher than those originally found in the water itself.

In recent years, much interest has been devoted to the process of eutrophication. This occurs when nutrient salts, which at appropriate levels are necessary for the growth of aquatic plants, are present in too great concentrations, so that there may be an undesirable growth of algae and other plants, and a decrease in the incidence of certain 'desirable' flowering plants.

A comparatively newly recognised problem affecting fresh water is acid rain. Certain lakes and rivers have been made more acidic by air pollution which has got into the rain, and which has been carried by the rainwater percolating through soils with a low buffering capacity into the streams and areas of open water. This has affected the whole economy of such water, exterminating fish and other aquatic organisms, with consequent effects on the populations of, for instance, birds which prey on fish and insects.

Pollution by organic matter has the effect of stimulating bacterial and fungal growth, and these processes absorb oxygen, and so de oxygenate the water. The organic matter is usually measured by the so-called biochemical oxygen demand (BOD). This is based on a test where the ability of polluted water to absorb oxygen is measured – the greater the oxygen demand, the more polluted the sample. In this test, a sample of contaminated water is incubated, in the dark, at 20°C for five days in a closed vessel containing a known amount of oxygen: the amount of oxygen taken up by the sample is its BOD. The higher the BOD, the more the amount of organic matter present, provided the same units are used. Thus we may have sewage, untreated, with a BOD of 300 mg per litre, farmyard wastes registering 2,000, effluents from paper pulp works of 25,000 mg 1, and silage effluent reaching an astronomical figure of 50,000 mg 1. As will be seen on p. 95, this has recently caused more pollution incidents in British rivers than any other substance.

The capacity of organic pollution to deoxygenate water is enormous. The faeces produced by a single human being gives rise to a daily oxygen demand of 115 gms. This represents the total oxygen which can dissolve in 10,000 litres of water. In a river like the Thames, with a flow which may be only 100,000,000 litres a day, complete de oxygenation could occur if the sewage from 100,000 people was discharged into it. This is what was happening 150 years ago.

Some mention was made in Chapter 1 of the way in which sewage has been dealt with over the centuries. Although earth privies have not completely disappeared from Britain, almost all houses are now supplied with water closets. In rural areas these may empty into septic tanks, or may even be allowed to drain away into porous soil. In towns various types of sewage works now exist; only on the coast is much raw sewage discharged, untreated, into the sea. Where there are long outfall pipes, running for a kilometre or more into deep water, this method has much to commend it. The sewage is diluted, and its only effect is to improve the nutrition of the fish and other marine life over a large area surrounding the outfall. What is unforgivable is that some towns still discharge their ordure from pipes which actually empty on the beach, or else only a few metres below low-tide mark. This has caused beaches to be contaminated, and is a health hazard to anyone unwise enough to bathe in the vicinity of the outfall. Such abominations have, in the majority of cases, been cleaned up, but some remain. Often the excuse (a quite genuine one) is that a proper inland sewage works is under construction, so it is not worthwhile undertaking costly improvements to an outfall which will only be used for a few more years.

Most urban sewage is treated by methods which essentially depend on oxidation by aerobic organisms. The most widely used system today includes filtration through trickling filters, which are the circular structures seen in most sewage works. They are made of clinker or broken stones, and the fluid trickles slowly through them, leaving the interstices full of air. The stones are covered with a mat of different micro-organisms, which feed on, and so remove, most of the organic matter. The filter is prevented from becoming clogged by the micro-organisms because insect larvae and worms develop in large numbers and feed on them. Another system of sewage treatment is the active sludge process, in which the sewage is run into tanks. These are inoculated with sludge from a previous batch, to ensure that the correct micro-organisms are present. To keep the tanks 'healthy' is an important and skilled operation. The contents are kept stirred to ensure efficient aeration. The organic matter is broken down, and a clear effluent, which may require further 'polishing' before it can be safely discharged into a river, is the result. The 'sewage sludge' which remains may be disposed of by dumping, for instance into the North Sea, a process which has given rise to international criticism although there is little evidence that it does any harm. The sludge may also be spread on the land as a fertiliser, either in its natural form, or after being partially 'dewatered' to yield filter cake which may be used in reclaiming derelict land (see p. 36).

The description of modern sewage plant is necessarily somewhat simplified. In many works the sewage is screened as it enters the system, to remove intractable objects and to break down the material. There are often tanks where considerable amounts of sludge are temporarily stored

90 WASTE AND POLLUTION

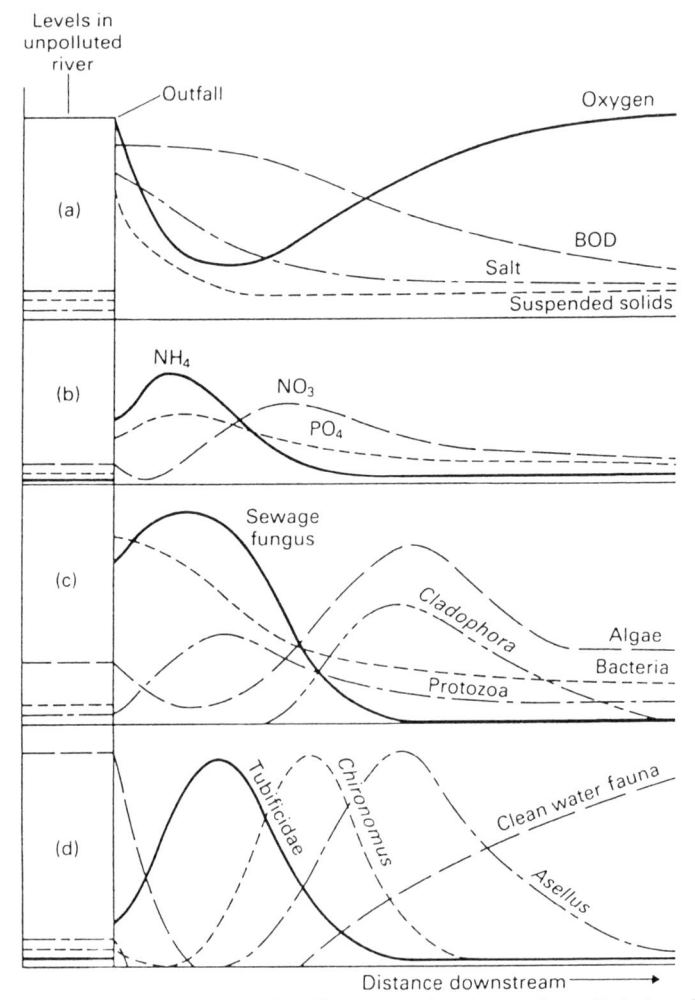

Fig. 12 The effects of an organic effluent on a river below the outfall. A and B physical and chemical changes, C changes in micro-organisms, D changes in larger invertebrates. (After H.B. Hynes)

before being disposed of; and there are still some lagoons where untreated sewage may be placed when the load on the machinery is, temporarily, too great. One trouble is that in cities storm water, i.e. the result of a deluge of rain, runs into the sewers and can virtually overwhelm the sewage works with a volume of liquid containing a comparatively small amount of organic matter. In some cases such contaminated storm water has to be discharged into a river or estuary which should only receive purified effluent. Emergency stores are being increasingly installed to avoid this unfortunate cause of intermittent pollution.

The modern sewage works is very efficient. When taking parties of visitors round, the engineers often demonstrate their confidence in the plant by drinking the effluent as it emerges. It apparently does them no harm, and we are assured that it tastes delicious. What is incontrovertible is that, diluted by the water of the river, the mixture is safe to drink. The water from the Thames which is abstracted to be drunk by Londoners is said already to have passed through five sets of human kidneys before emerging from the London taps. This is, of course, an exaggeration. Only a tiny fraction of what is drunk has been subject to these treatments, but nevertheless the safety of the water is a tribute to the skill of our engineers.

When an organic effluent, be it untreated or partially treated sewage, or the run from an intensive animal holding, enters a river, it has a profound effect, as illustrated by Fig. 12. The top section shows how the oxygen level falls, and then rises again downriver where the self-purification of the river has taken place, and also where further atmospheric oxygen has dissolved. The BOD falls as the oxygen level rises. The second section shows the chemical changes, the third the way in which micro-organisms develop. Finally, the bottom section shows what happens to the larger invertebrates. The process of self-purification is important. With a small amount of pollution and a large river, it is rapid and the damaged stretch of the river is short. Unfortunately some rivers have been so heavily polluted that the self-purification process cannot be effective. In such cases it will take many years, with efficient treatment of the sewage, before the river is restored.

Different invertebrate animals are found where the pollution levels vary. It has, in fact, been possible to use different species as indicators of varying degrees of pollution. In 1971 an exercise was mounted by the Advisory Centre for Education through the *Sunday Times* newspaper. A kit was prepared, showing the typical invertebrates found in various freshwater sites with different amounts of pollution (Fig. 13). Thus Stone Flies and Mayflies were described as being restricted to the cleanest water, Caddis Fly larvae and Freshwater Shrimps to the next cleanest, Water Lice and Bloodworms to more polluted water, Sludge Worms and Rat-tailed Maggots to quite seriously polluted sites, with a final category where pollution prevented any apparent life. The exercise, which included filling in the form (Fig. 14) was performed by some 8,000 children, with an age range indicated in Fig. 15. As one of those who analysed the results, I was surprised by the high standard of the

stonefly nymph

brownish colour

two long tails

no gills on abdomen; there may be gills under the thorax where the legs arise

size you might expect

six jointed legs

long antennae

HEAD

A

A crawler, stonefly nymphs are very slow moving. Found in clean running water with a gravel or stoney bed. The nymph stage may last four years with frequent moults. Most stonefly nymphs are plant eaters; a few eat other kinds of nymph.

The photograph is of *Perlodes*

mayfly nymph

brown mottled colour

size you might e[xpect]

antennae

gills along the abdomen; you can see them moving (except in Caenid mayfly)

HEAD

three long thin tails

six jointed legs

B

A swimmer. Usually moves quickly by beating its abdomen *up and down*. The *damselfly*, which might look similar at first glance, swims with a side-to-side movement, rather like a fish. The damselfly does not have gills along the abdomen.

Mayfly nymphs are found in a wide range [of] conditions – both fast water and muddy be[d]. The nymph stage may last for three years, moulting frequently – hence the old 'skins' quite common. The adult, which might on[ly live] a few hours, doesn't feed. The nymph is pl[ant] eating.

Note: some mayflies (the Caenids) have gil[l] covers protecting the gills. This makes the [m] hard to see.

Photograph of *Baetis sp.*

caddis fly larva

usually lives in a case – the case made of plant matter, sand, or stones.

six legs

size you might expect including case

C

head often brightly coloured – yellow and brown, yellow and black, green

hooks at end of abdomen (you won't see these if the insect is in a case)

Although most caddis larvae live in cases, not all do. The case is usually **carried around**, when the **insect crawls**. One type does swim though, and sometimes the case is fixed to a plant or stone. Between this larva and the adult there is another stage in the life cycle – the pupa (= chrysalis of a butterfly). Then the case is attached to a plant or stone, and the ends of the case closed off. The larvae are usually plant eaters.

Photograph of *Phryganea sp.*

freshwater shrimp

body flattened sideways

grey-brown colour

looks like a sea-side shrimp

HEAD

D

many more than six jointed 'legs' – and all the way down the abdomen too

long antennae

size you might expect

Swims – quickly and **often on its side**. It can al[so] crawl. Usually found with fastish flow. The eg[gs] are carried in a pouch and the young look lik[e the] adult. Shrimps feed on dead and decaying ma[tter] (sometimes live prey).

Photograph of *Gammarus sp.*

water louse

body flattened – wide flat body

grey brown colour

HEAD

long antennae

many jointed 'legs'

looks rather like a wood-louse

size you might expect

E

Crawls, does not swim. The eggs hatch straight into small adults. Water lice feed on dead and decaying matter, especially plant material.

The photograph is a view from above of *Asellus sp.*

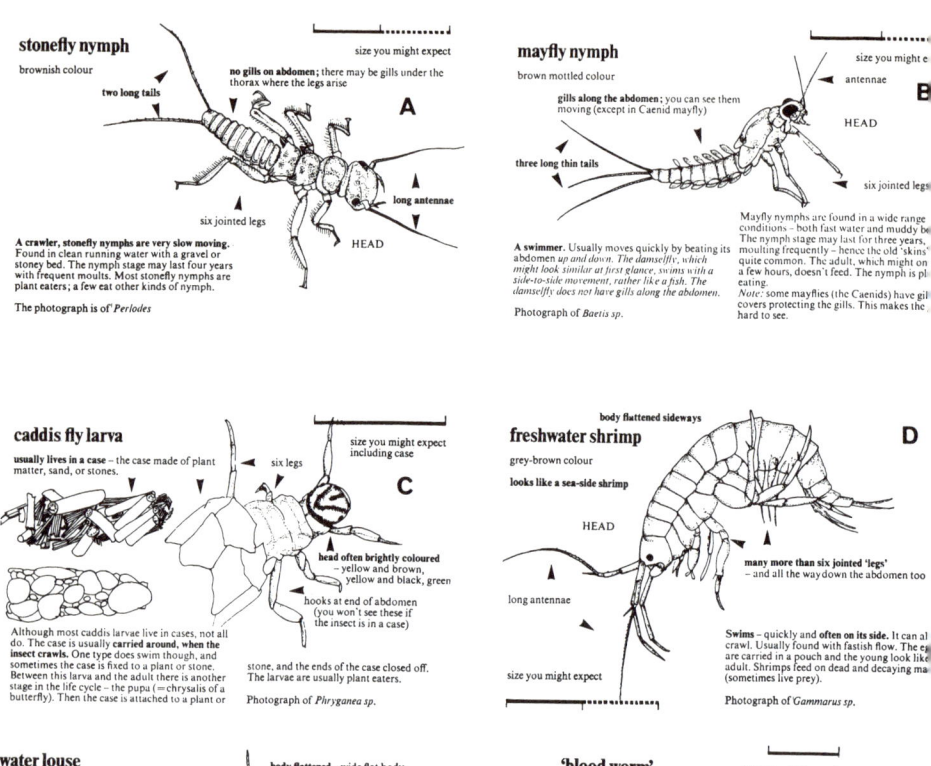

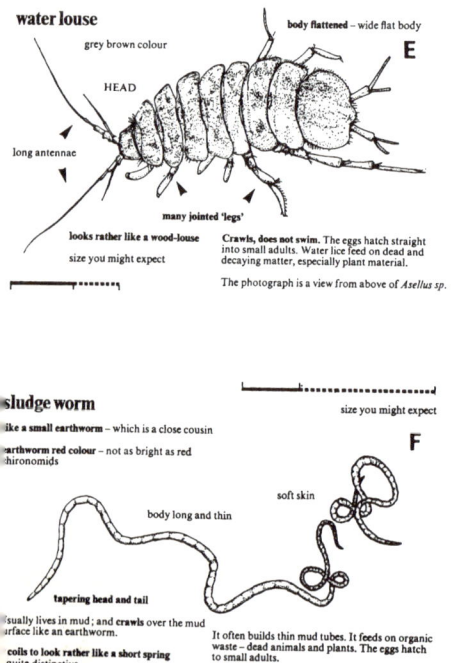

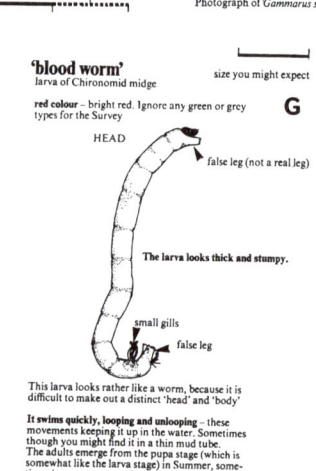

'blood worm'

larva of Chironomid midge

red colour – bright red. Ignore any green or grey types for the Survey

size you might expect

HEAD

G

false leg (not a real leg)

The larva looks thick and stumpy.

small gills

false leg

This larva looks rather like a worm, because it is difficult to make out a distinct 'head' and 'body'

It swims quickly, looping and unlooping – these movements keeping it up in the water. Sometimes though you might find it in a thin mud tube. The adults emerge from the pupa stage (which is somewhat like the larva stage) in Summer, sometimes in enormous numbers.

Photograph of *Chironomus sp.*

sludge worm

[l]ike a small earthworm – which is a close cousin

[e]arthworm red colour – not as bright as red [c]hironomids

size you might expect

F

soft skin

body long and thin

tapering head and tail

[u]sually lives in mud; and **crawls** over the mud [s]urface like an earthworm.

coils to look rather like a short spring quite distinctive.

It often builds thin mud tubes. It feeds on organic waste – dead animals and plants. The eggs hatch to small adults.

Fig. 13 Descriptive diagrams of indicator animals

Here is a survey report form you might use and copy. It will help you when you compare different streams and rivers, or when you investigate the same stream at different places or at different times of the year.

2 The river or stream
A short stretch of running water – stream or river, town or country. Make sure the banks are safe, and that you're not on private property without permission.

Only fill the map reference in if you know it accurately.

tick

4 pH *use some of your Universal Indicator paper*

5 Indicator animals
Try, if you have time, to carry out more than one trial. Do three trials a few yards from each other. Try to find similar flow conditions each time.

beware of the damselfly (see the broadsheet picture)

tick only if you have a red one

finds of A indicate clean water. B C D may also be present
B can withstand some more pollution than A
finds of C D only indicate pollution of a fairly serious nature
finds of D only (or E of course) suggest highly polluted water

6 Fish
If people are fishing near the spot where you are investigating try to find out what fish they have caught, or expect to catch.

Survey report

1 your name   age

– home address or school address

2 name of river or stream being investigated

– nearest village or town   county

– Ordnance survey map grid reference
  of spot being investigated

3 date of survey

– has there been, within the previous three days
  heavy rain   some rain   no rain

4 pH of water in the river or stream

5 Indicator animals

Place a tick opposite the species you found–

	first trial	second trial	third trial	
stone fly nymph	☐	☐	☐	A
mayfly nymph	☐	☐	☐	
caddis fly larva	☐	☐	☐	B
freshwater shrimp	☐	☐	☐	
water louse	☐	☐	☐	C
'blood worm' (chironomid larva)	☐	☐	☐	
sludge worm and only those breathing air, eg rat-tailed maggot	☐	☐	☐	D
no apparent animal life	☐	☐	☐	E

6 names of fish being caught

This pack is devised by:
Things of Science (Pollution Inquiry)
Advisory Centre for Education
32 Trumpington Street
Cambridge
to whom all correspondence or queries should be addressed

Shenval

Fig. 14 Survey report form

94 WASTE AND POLLUTION

Fig. 15 Age of participants in water pollution survey (After Mellanby 1974)

returns, even those from the youngest children. The survey was the most detailed which had ever been made in Britain. Where results of pollution from 'official' sources existed, the children's survey data agreed.

The survey had some practical results. One small boy was horrified to observe that the effluent from a local laundry, entering a relatively clean river with good populations of Caddis and Freshwater Shrimp, transformed it so that these species disappeared and, at least for a distance, little life remained. The child marched into the laundry and demanded to see the manager, and told him of his findings in no uncertain terms. The manager called him a silly little boy and told him to b——r off. So the boy went straight to the local newspaper, which reported his findings. The unfortunate manager had to take immediate steps to deal with the effluent. What I like about this story is that the boy was able to act on hard evidence, something often lacking in the attacks on industry by certain environmentalists and by the journalists supporting them.

Fish have, of course, long been considered to be indicators of pollution. Before modern legislation controlling pollution had been enacted, organisations like the Anglers' Cooperative had relied on the common law, which gave the right of riparian owners to enjoy pure water passing along their banks. In many cases they successfully prosecuted polluters, and did much to preserve the purity of the rivers they fished. Difficulties arose where a river was grossly polluted from a great number of sources, and where no individual could be shown to be mainly to blame.

As indicated in Chapter 1, agricultural wastes amount to some 200,000 million tonnes. Much of this is slurry, which resembles human sewage except that it is produced in much greater amounts. A great deal is used as manure, and is thus recycled through crops. Many farm animals, however, are kept intensively by farmers with no land on which to grow crops and thus utilise their manure. They then run it into lagoons where it oxidises, or on to limited areas of land where it also breaks down. Unfortunately a great many accidents occur, and pollution from the run-off from cattle yards and slurry storage is common. There were 648 such incidents in 1984, and, although only 46 led to successful convictions, most farmers with livestock are now being more careful. Some slurry is used to generate methane, which is burned on the farm. A number of farmers have built quite sophisticated sewage treatment plants.

Silage effluent, as mentioned, has the highest BOD on record. Again in 1984 there were 398 serious incidents when it gave rise to pollution, and 32 cases of successful prosecutions. As the volumes of this liquid are relatively small, it should not be difficult to prevent this damage from occurring in the future, if all silage clamps are properly designed and sited so that the effluent is either treated or permanently contained.

Most of our rivers today are getting cleaner rather than dirtier, though we still have trouble from old and overloaded sewage works. The most often quoted success story concerns the Thames. In the days of the Stuart kings, this was an excellent Salmon river, and apparently Charles II had a pet polar bear that used to catch Salmon, to the delight of the onlookers. As London's population rose, more and more of its dirt entered the river. As stated in Chapter 1, this reached its climax when water closets were introduced in the 19th century, and all the sewage was discharged untreated into the Thames. The situation was exacerbated in that much of the foul water remained within the London area for days, being pushed up the river by rising tides, and then coming only a little further down when the tide ebbed. Sewage treatment was introduced, but the improvement was gradual, and even by the 1930s the river was fishless and frequently quite devoid of any oxygen. Since then things have greatly improved. The Thames now supports no less than 110 species of fish, and this includes Salmon which are notoriously susceptible to damage from pollution. This return of fish has encouraged the return of fish-eating birds. Thus the low- tide mud between Woolwich Ferry and Margaret Ness now supports 600 Pochard and 610 Mallard. The bay to the east of Barking Power Station is an important site. Here counts have revealed 326 Mallard, 630 Teal (*Anas crecca*), 95 Pintail (*Anas acuta*), 100 Tufted Duck (*Athya fuligula*), 1,500 Pochard (*Athya ferina*), 800 Shelduck (*Tadorna tadorna*), 30 Dunlin (*Calidris alpina*) and 5 Redshank (*Tringa totanus*). These results are quoted as characteristic of the improvement which has now occurred. Similar or even larger counts have been made all down the Thames, where until recently few such birds were to be seen.

Gross organic pollution has therefore been largely conquered, and there is no reason why all rivers flowing through urban or industrial areas should not be as clean as the Thames within a few years. Unfortunately there will still be problems. The Mersey, formerly one of the most polluted rivers, is an interesting case. In 1979, at a time when we thought things were significantly improved, there was a serious mortality of birds, and several thousand perished. The species worst affected was the Dunlin, but Teal, Herring Gulls (*Larus argentatus*) and Black-headed Gulls (*Larus ridibundus*) also suffered. The cause of death was diagnosed as lead poisoning. The source of the lead was the firm Associated Octel, who manufactured tetra-ethyl lead used in petrol to raise the octane rating. Ironically, although this firm had discharged its effluents, quite legally, into the Mersey for many years, when the bird deaths occurred the situation had greatly improved and discharges had been largely curtailed.

The final explanation of the bird deaths was interesting. While the Mersey was grossly polluted, few birds were seen on the mudflats, as these contained very little invertebrate life on which the birds could feed. As the river became less polluted, molluscs and other invertebrates began to colonise the mudflats and built up considerable populations, which attracted the Dunlin and other birds. The mud had, over many years, built up a considerable concentration of alkyl lead, a substance which is very much more poisonous than any inorganic lead compound. However, the levels of alkyl lead were not sufficient to poison the invertebrates. The birds, at the end of the food chain, apparently concentrated the lead compounds up to lethal levels, hence the deaths. This is an instance where an environmental improvement, i.e. the reduction of pollution levels, caused an increase in bird mortality, at least for a period. It is difficult to see how this could have been avoided.

Scientific literature contains many examples of the way in which poisons in very dilute concentrations in water have been taken up by fish, which have developed high levels, sometimes lethal levels, in their bodies. In one instance a load of the insecticide endosulphan was discharged into the river Rhine, killing vast numbers of fish. The authorities in the Netherlands, where Rhine water was abstracted for drinking, stopped this immediately, for fear of endangering the human population. This would have been very unlikely. It was calculated that a person would have had to drink a million gallons of water to ingest a harmful dose, and the rate of excretion would in fact have exceeded the rate of absorption. The fish, however, had to 'drink' such large volumes (see p. 87) that they were exposed to lethal amounts.

This process of poison concentration by aquatic animals is important.

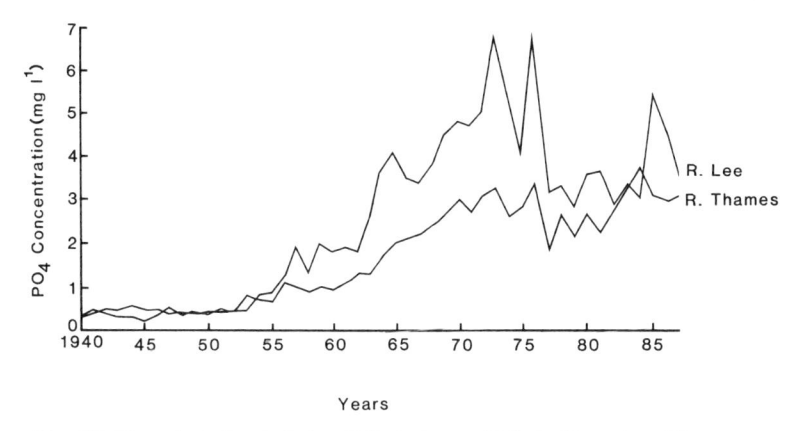

Fig. 16 Phosphate levels in two British rivers, 1940-8. (Data from Thames Water Authority)

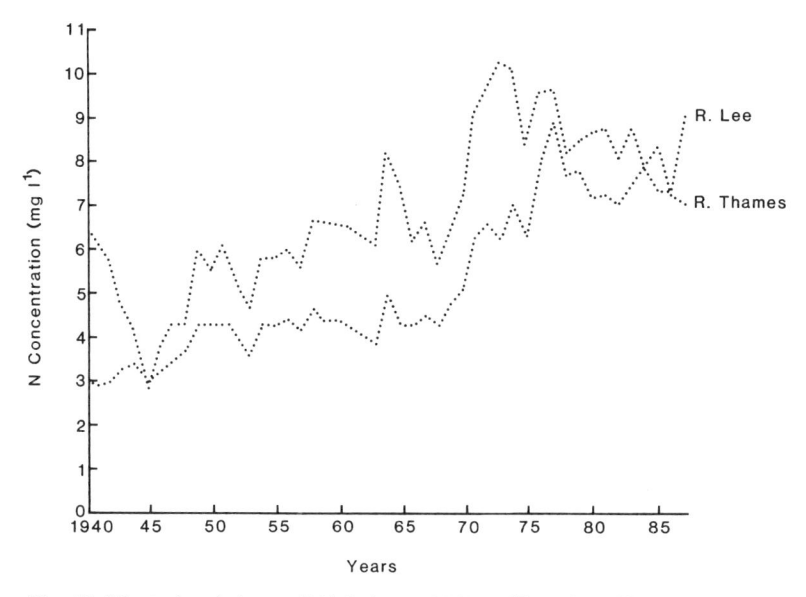

Fig. 17 Nitrate levels in two British rivers, 1940-8. (Data from Thames Water Authority)

As in the case of the endosulphan, it is the fish (or other animal) which receives the largest dose. Subsequent concentration up various stages of a food chain, which has received more publicity, is probably less important, though it does occasionally happen.

Sewage treatment works remove the organic matter, but they discharge most of the plant nutrients, particularly nitrates and phosphates, into the rivers. This has caused a considerable rise of phosphate and nitrate levels in the waters which receive the emissions. Fig. 16 shows that the Thames had an increase in phosphate from less than half a milligram per litre in 1940 to around 4 milligrams in 1975. This level, with some fluctuations, has continued. The river Lee, which flows into the lower Thames, increased its phosphate level to about 8 milligrams (twice the level found in the Thames) in 1975. Here again there has been no substantial increase since 1975. A somewhat similar result has been obtained with nitrate. The levels rose substantially until 1975, and then remained fairly constant, as shown in Fig. 17. It is interesting to note this result for nitrate. As will be explained, there has recently been much concern regarding the possible harmful effects of nitrates on human health. The fact that levels have ceased to rise at a time when the use of nitrogenous fertilisers has continued to increase suggests that these substances may not contribute so much as has been suggested to the levels of nitrates in drinking water.

The levels of phosphates and nitrates in our rivers are nevertheless generally much higher than they were 40 years ago, and this is affecting the living organisms in the rivers. Where the water is flowing, changes in the vegetation may not be easily detected, but the results of eutrophication are considerable when the water is impounded in lakes or reservoirs. Many people, when speaking of this phenomenon, tend to concentrate on the increase in the level of nitrate. This is important as it affects human beings, but most waters, even when fairly low in nitrates, contain sufficient to support increased plant growth if phosphate, which is usually in short supply, is added.

Nitrate levels in water depend mainly on leaching from the soil. Little phosphate, even when large amounts contained in fertilisers are applied to the soil, is leached out in this manner. Increased phosphate usually comes from sewage effluent. The high levels found today are derived from the increased use of phosphate-rich detergents included in modern, efficient washing powders, which go down the drains and are then discharged in the effluent. In some lakes another important source is the large number of roosting gulls which pass their phosphate-rich excrement into the water. Reservoirs fouled in this way may also be polluted with *Salmonella*. One survey showed 52 out of 111 samples from a Scottish loch so infected, probably by gulls transporting the bacteria from a refuse tip.

Still water with a high phosphate level soon develops a substantial algal growth. This may be kept in check if there is a healthy population of small crustaceans, such as *Daphnia* sp., to graze upon it. In one case a reservoir near Bristol became covered with algae. Chemical analyses showed that the levels of nutrients had not changed. Eventually the cause was tracked down to waste sheep-dip which had been dumped in a hole in the ground several miles away. This had slowly seeped through the intervening soil, and a quantity of dieldrin sufficient to kill off the crustaceans had reached the Chew Valley lakes. With the grazers gone, the algae took over. This alga, here and in other lakes, casts shadows in deeper water and inhibits plant growth. When the alga dies, the rotting organic material will often deoxygenate the water which, if drunk, has an unpleasant flavour. So eutrophication, generally triggered by rises in phosphate, may have undesirable biological effects.

One area where eutrophication, and in particular increased phosphate levels, has caused ecological deterioration is the Norfolk Broads. This is an area where medieval peat cutting has produced a series of 'broads', shallow expanses of open water connected by rivers and channels, with a rich marshland vegetation surrounding the water. This is a popular holiday area, with a very large number of motor cruisers and other craft. It is also greatly prized for its vegetation, wildlife, including insects and birds, and fishing. In recent years, however, much of the vegetation has disappeared, the previously clear water has become cloudy, the area of reedbeds has greatly diminished, and fishing has become less productive.

Tourist pressure has been blamed for some of the damage. Motor cruisers driven too fast cause a damaging wash, and litter and other rubbish from boats pollutes the water. At one time the boats emptied their sewage untreated into the water, but this has been stopped. It is now believed that, with care and cooperation from all concerned, the tourist trade can be tolerated provided it does not increase too much.

There has been much research to discover the cause of the trouble, and to find ways in which the harmful trends can be reversed. B.Moss, of the University of East Anglia, has contributed considerably to our knowledge of the situation. The main reason for the deterioration of the Broads is undoubtedly increased eutrophication, particularly the rise in phosphate levels in the water. In many parts of the Broads, the submerged macrophytes have been replaced by phytoplankton-dominated communities. The loss of submerged plants appears to be linked with decreased invertebrate diversity, so that the algae are not kept in check by grazing. Eutrophication is also associated with fish-kills caused by a toxic flagellate *Prymnesium parvum*, outbreaks of avian botulism and decreased diversity of fenland plants and animals through flooding with enriched water.

Restoration has been attempted, mainly by the reduction of phosphate

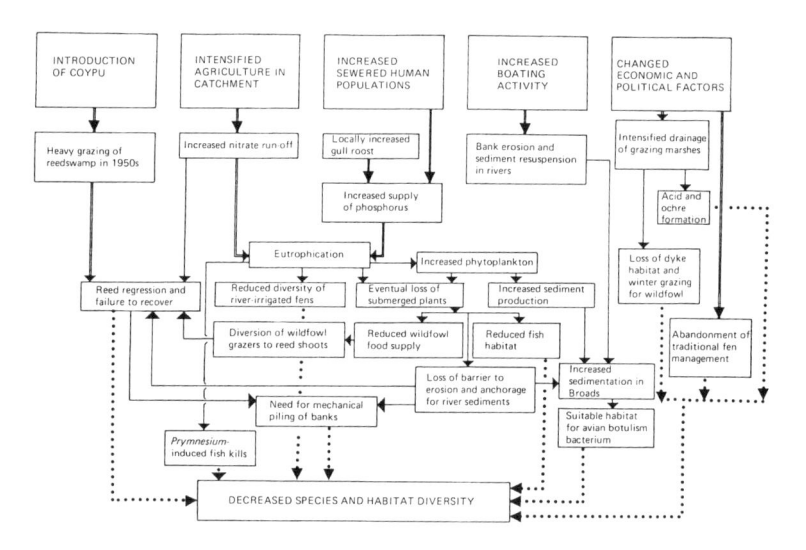

Fig. 18 Interrelationships of the Broadland problems. (After Moss 1987)

levels and phosphate input. The difficulty is that phosphate is released from the sediments for a long time even if input of phosphate is substantially reduced. Eventual success, however, seems probable. The Anglian Water Authority has cooperated by introducing 'phosphate stripping' at one of its main sewage treatment works at Stalham. This reduces phosphate output by nearly 90 per cent. Attempts have also been made to remove phosphate-rich sediment and dump them on the land where they improve the fertility. Some areas of open water have been isolated so that no more phosphate enters, and here levels have fallen, but this solution cannot be used where navigation is permitted. It is still too early to predict the likely success, in the long run, of these different methods, but results are promising. The phosphate stripping technique is being widely tried out, and is likely to be introduced in other parts of the country. The whole system is summarised in Fig. 18, taken from Moss's publications.

We have recently heard much about the danger to human health of the rise in nitrate levels in drinking water. The usual complaint is that farmers are using too much chemical fertiliser, and that much of this is finding its way into rivers, reservoirs and underground aquifers from which drinking water is derived. The levels of nitrate in some drinking water, particularly in East Anglia where intensive arable farming uses the greatest amount of fertiliser, from time to time exceeds the EEC maximum level of 50 mg nitrate per litre. The reason for the concern is that high levels of nitrate, if used to make up bottle feeds of young infants, may cause sickness, the nitrate being turned to nitrite which combines with the child's haemoglobin and produces the 'blue baby' syndrome. There have also been suggestions that the raised levels of nitrate may cause stomach cancer, but this is not supported by recent research. If nitrate is harmful to humans, it is also likely to damage wild animals, but I have seen no data to support such a supposition. In Britain there have, in fact, been no cases of blue babies caused by drinking water from the tap, but it is obviously wise to act preventively. The highest levels of nitrate have been found in farmyard wells where contamination from manure occurs.

Today there is a call for controls on the amount of nitrogen which farmers may use to fertilise their land. This is hardly a new idea. In an article I wrote called 'Hidden Dangers in Modern Farming Chemicals' which appeared in *The Times* on 2 July 1969, I said, on the subject of rising nitrogen levels in drinking water: 'Strict regulations on fertiliser use might have to be enforced in catchment areas.' Now I am not sure that this would be effective, for reasons stated below.

The cause of raised nitrate levels is much more complicated than the above-mentioned views might suggest. The most worrying levels are in water from aquifers which are deep below ground level in the chalk. They are recharged by water slowly percolating down to them. This water moves very slowly, and only now may we be using water which came down as rain half a century ago. At that time, the use of nitrogenous fertilisers had scarcely started to increase. So there seems to be no direct correlation with application to crops and levels in the aquifer. This discovery is not very reassuring, as it suggests that much higher nitrate levels in the aquifers may be on the way, and there is nothing we can do to stop this happening, at least for a period until reduced surface levels have been ensured.

The important thing is to find the source of the nitrate leaching out of the soil. In fact, the nitrate applied to winter wheat, in doses up to 200 kg per hectare, is almost entirely absorbed by the crop if applied at the right time. At harvest, there is little free nitrate in the soil. The loss from arable fields takes place mainly in autumn and early winter, when the fields are bare or only partially covered with growing crops, and is the result of bacterial action on the nitrogen locked away in the soil's organic matter. The highest losses may come from organic farmers who apply heavy dressings of farmyard manure in autumn, when the greater part of the nitrogen is lost before it can be used by any crop. All sources of soil nitrate can be lost by bacterial action, and can contribute to levels in water supplies. It does not matter whether it is produced by legumes, or whether it is stored in the turf of grass leys. If these are ploughed in autumn, much of the nitrate literally goes down the drain.

So simply to pass laws to reduce the use of fertilisers, or to put a punitive tax on them, might do little to solve the problem. Organic farming might prove no better. A 'code of good farming practice', aimed at reducing the nitrate running off our fields, makes the following suggestions:

1. Do not apply nitrogen fertiliser or farmyard manure in the autumn.
2. Do not leave the soil bare during the winter.
3. Sow winter crops early in the autumn, to get a good soil cover.
4. If a spring crop is to be sown, grow a winter catch crop.
5. Use animal manures judiciously.
6. Do not plough up too much grassland at one time.
7. Plough in straw. This locks up some nitrate (and may reduce yields in the first year and cause severe infestation by slugs but may give increased nitrate production in the long term).
8. Use nitrogen fertiliser in accordance with professional advice. Apply it only when the crop is growing actively.

These recommendations should improve the situation, but will not prevent the occurrence of levels above the EEC recommendations. It is perhaps reassuring to find there is no evidence that levels twice or more than twice the 50 mg litre set by the EC do any harm to adults, even those who drink a great deal of water. There is little reason on these grounds to incur the considerable expense of using bottled mineral water.

The suggestion, incidentally, that because most of the nitrate is lost from the soil in winter, this will inhibit algal growth, as the temperature is too low, is based on a misunderstanding. Nitrate alone, at any time of year, only encourages algal growth when there is sufficient phosphate present, so as far as this undesirable effect of eutrophication is concerned, nitrate loss from the soil is largely irrelevant.

Many writers on natural history give positively lyrical accounts of the excellence of 'sewage farms' as sites for the observation of birds, particularly migratory waders in winter. These so-called farms were of various types. Some were truly farms, where untreated sewage ran along furrows with vegetables growing on the ridges. Cabbages and other plants grew plentifully, and no doubt provided excellent eating provided cooking killed any harmful parasites. Most sewage works did little genuine farming, but they included large lagoons where the raw sewage matured and oxidised. In many cases the lagoons were allowed to dry out from time to time, and occasionally the area was sown with crops. As a rule it was left to develop its own vegetation unaided. Many ruderal species grew most luxuriantly and produced good cover and ample amounts of seed encouraging huge populations of finches. Species whose seed survived in the sewage, like tomatoes, were common, and in good summer yielded reasonable crops.

This type of sewage farm encouraged bird life. In many cases the fairly large sites included considerable areas of trees and shrubs which provided nesting sites for certain species. The data for Rye Meads Sewage Purification Works in Hertfordshire, for the period 1957–62 are given as Appendix 3. This shows that 52 species bred on the site, with a total of 300 breeding pairs, not including the numerous House Sparrows; 58 non-breeding visiting species were recorded on more than 10 occasions, and 57 vagrant species were seen on less than 10 occasions. The breeding pairs included 34 Sedge Warblers and 30 Reed Buntings. It is tempting to dwell on the many other similar sites, and describe the waders, including many rarities, which came there during their winter migration. In 1938 and 1939 these species at a London site included Ringed and Golden Plover (*Charadrius hiaticula* and *Pluvialis apricaria*), Ruff (*Philomachus pugnax*), Dunlin, Common and Green Sandpipers (*Tringa hypoleucos* and *Tringa ochropus*), Greenshank *(Tringa nebularia)*, Curlew (*Numenius arquata*), Whimbrel (*Numerius phalopus*), Jack Snipe (*Lymnocryptes minutus*), and rarities including the Avocet (*Recurvirostris avosetta*) and Temminck's Stint (*Calidris temminckii*). Unfortunately none of these sites remain. Their obituary appeared in an article by A.W. Boyd in *British Birds* in 1957, who said, 'Such then are the conditions of the sewage-farms which before long will be only a memory, as modern methods of sewage-disposal take the place of that which for 50 years has served the birds so well.' What is perhaps surprising is that there was no protest from ornithologists or conservationists when these sites were destroyed. This contrasts with what happens today. When a semi- derelict reservoir, from which most of the water has been removed, at Kempton Park was proposed for removal in 1987, the Nature Conservancy Council, with the support of the London Wildlife Trust, proposed to declare it an SSSI on account of the handful of birds which were visiting it. Perhaps the smell from the lagoons, which in warm weather with the wind in the wrong direction made life unpleasant for some miles from the plant, had its effect on public opinion. The modern sewage works provides few facilities for birds. Psychodid midges breed in the filters, and Wagtails and Starlings frequent these in some numbers. Where there are temporary facilities for storing sludge, a few waders may settle for a short time, but never in anything like the numbers, or the variety of species, found on the old type of sewage farm.

The subject of acid rain and the way this affects terrestrial habitats, was discussed in Chapter 6 (p. 78). We saw that the situation was very complicated, involving many factors, including climate, pathogens and nutrition as well as pollution. A clearer picture can be drawn regarding the effects of acid rain on freshwater. There is no doubt that lakes and rivers in Norway, Sweden, Canada and Britain have, in recent years, become much more acidic than they were previously, and also that fish, particularly Salmon and Trout, have disappeared from some of these waters. There is reasonably good evidence to suggest that the most important cause of these changes has been acid, much of it in the rain, and that this acid has arisen from waste gases containing sulphur and nitrogen produced by power stations in many countries and also from exhaust gases from motor vehicles.

The first important point to note is that lakes and rivers have only become acidic in areas where the rocks, and the soil derived from these rocks, contains little calcium, and where the soil has little buffering capacity. The presence of calcium in the water has also proved to be important – moderately acid water with reasonable levels of calcium is able to support fish which would not be present at the same pH if calcium levels were very low.

Fig. 19 shows the areas in Britain where the rocks and soil are such that acid deposition is likely to have a significant effect on the groundwater. The same diagram shows the contours for the mean annual pH of the rain. It will be seen that in Galloway, Scotland, and in Wales, where the effects of increased acidity have been observed, the rain is actually less acid than in the east of England, where no such effects have been found. This is partly due to the geology of these areas, but also to the differences in rainfall. Whereas pH is a useful measure of acidity, what matters is the total amount of acid deposited. Because the above-mentioned areas have the heaviest rainfall, they show comparatively large acid deposits. The same is true of Norway, where the mountains attract heavy rain.

Land use is also important in areas where water acidity may be unfavourable. The planting of coniferous forests can substantially increase the acidity of the water which drains from the afforested area. The reasons for this are complex.

Conifers produce acidic litter, but probably the most important cause of the increased acidity is that the trees 'scavenge' dry deposition of sulphur dioxide and other substances from the air, and this is washed off the trees by the rain. Even if the air is sufficiently low in sulphur dioxide to allow the growth of shrubby lichens (see p. 76 above), a significant amount is, in time, deposited on the trees, and eventually can build up substantially in the soil.

Increased acid in water has profound biological consequences, not

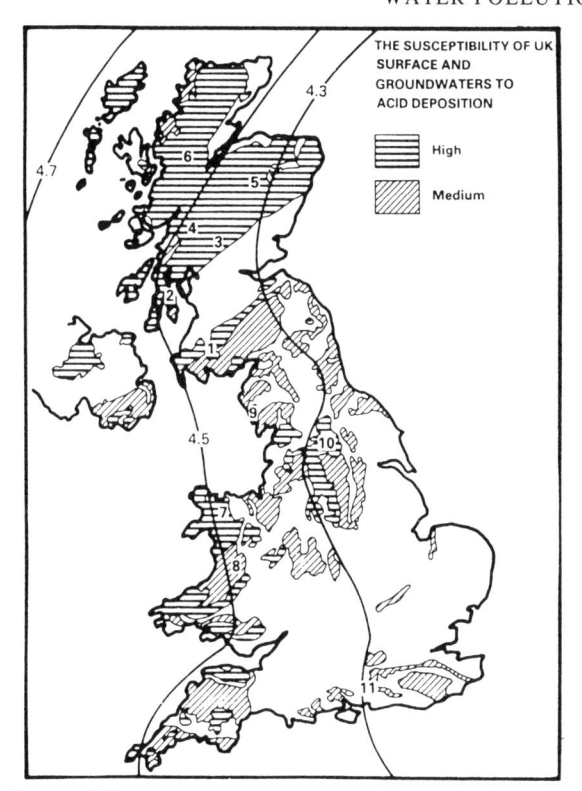

Fig. 19 Susceptibility of UK surface and groundwaters to acid deposition, and contours of annual mean pH of precipitation (After Mellanby 1988)

only on fish but also on plants and invertebrates. The fish are, in effect, poisoned by too much acid, and they may also be affected by the higher levels of aluminium which the acid washes out of the soil. The populations of the aquatic invertebrates are also generally reduced by increased acidity, and this can reduce the food available for birds and fish which feed on them. One case which has been studied is that of the Dipper (*Cinclus cinclus*), whose numbers in Wales have decreased significantly as the streams they haunt have become more acid, because of reduction in the insects on which the Dippers feed. Ospreys (*Pandion haliaetus*), which became extinct in Britain in the 19th century, have shown a welcome return, and in Scotland, when not harried by egg collectors, they have flourished, the population has increased, and breeding has been satisfactory. However, it has been suggested that more areas in Scotland and Wales would have Ospreys were it not for the reduction in fish stocks due to acidification of the waters.

The results of acidification on wildlife have, therefore, been serious. Yet all is not lost. The fall in the output of sulphur in Britain has been very significant. R.W. Batterbee, who pioneered the work on diatoms in lake sediments which established the way in which acidification had occurred during the last 150 years (since the Industrial Revolution) and which showed that this process had accelerated since about 1960, has very recently noted some reversal of acidification in the Galloway lakes, with changes in the diatoms to support this observation; and in Sweden and Canada similar observations have been made. These findings provide good evidence, incidentally, that the acidity was, in the first instances, related to the sulphur output, underlining the fact that attempts to reduce the output are likely to be worthwhile.

The effects of acidity can be neutralised by adding lime to the water, and this also has the advantage of raising the calcium levels. In Scandinavia lime or limestone has been added to lakes and rivers, generally with beneficial results. The pH has been raised, and in many cases fish have returned. However, the process of liming has to be repeated yearly in lakes and even more frequently in fast-running rivers. Most ecologists consider it to be only a temporary measure until acid deposition is controlled; but, as has been pointed out, although the procedure is expensive, the costs are generally much less than those of reducing the output from power stations.

The whole question of the control of sulphur output is complicated. The Norwegians, in particular, have blamed the British Central Electricity Generating Board for being the major cause of acidification of their rivers. Actually, only between 10 and 20 per cent of the acid deposition in Norway comes from Britain. Since 1970 Britain's output of sulphur has fallen by 40 per cent, yet until recently there has been increasing damage to Norwegian rivers. This did not encourage the CEGB to spend some hundreds of millions of pounds on treating their flue gases when the results might well have been negligible. The decision has now been taken to reduce Britain's output of sulphur considerably, and although this may not be for a very good scientific reason, it is probably a good political gesture. It has also been shown that the output in Britain is probably having its maximum effect within this country.

The manner in which acidification has occurred, and the possible way in which the damage can be cured, is demonstrated by the story of Loch Fleet in Galloway in southern Scotland. This loch, 17 hectares in extent, has a well-defined catchment area of 111 hectares. Up to 1950 it provided significant catches of Trout, but these declined and after 1965 only the occasional fish was caught (Fig. 21). There are no records of the acidity during the early years, but measurements after the fish had gone were in the region of pH 4.5 – 5. Instead of adding lime of limestone to

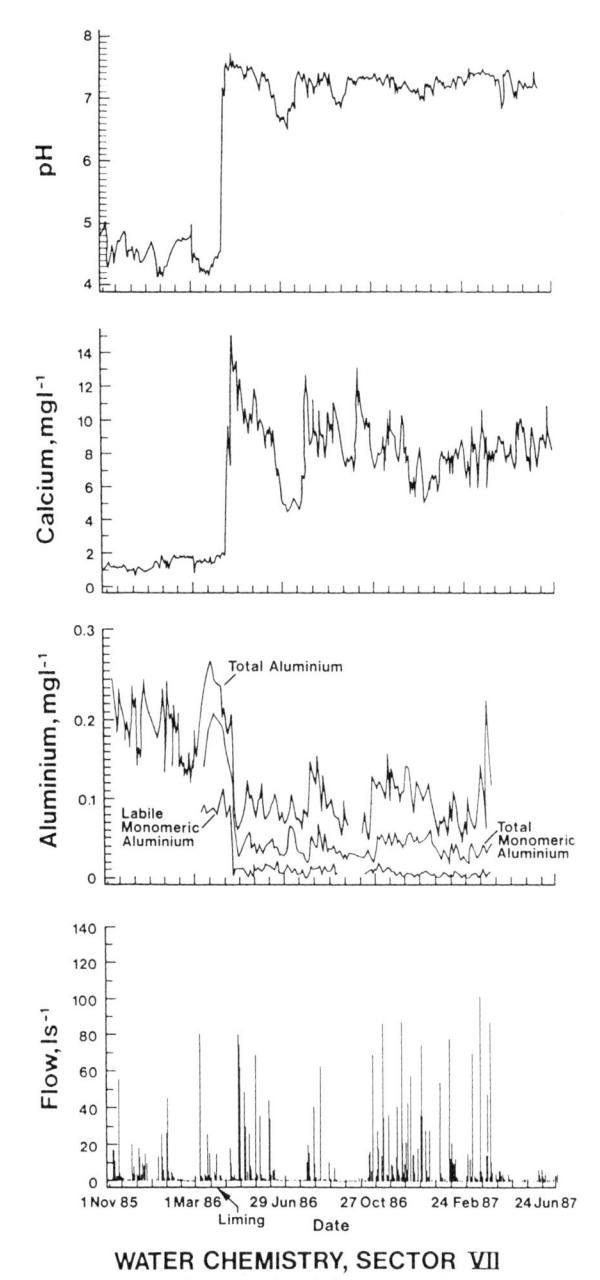

Fig. 20 Loch Fleet water chemistry before and after liming the catchment area (CEGB 1987)

BROWN TROUT CAUGHT SINCE 1935

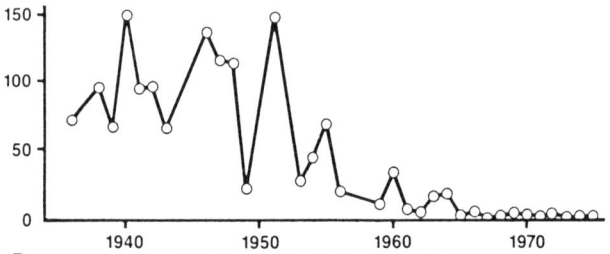

Fig. 21 Brown trout caught in Loch Fleet since 1935. (CEGB 1987)

the water, 300 tonnes of limestone were applied to the catchment in April 1986. As Fig. 20 shows, there was an immediate response. The pH rose to 7.6, the calcium from under 3 to over 10 mg 1, and the aluminium levels were reduced by some two-thirds. The water of this lake is replaced approximately once every six months. Results so far suggest that the treatment of the catchment should keep the pH and the calcium levels up to values suitable for Trout for perhaps ten years – much more effective than applying lime to the water annually or even more often. Trout certainly survive in the loch, and it seems likely that its value for wildlife of all kinds will be fully restored.

The general conclusion regarding the effects of acid rain would seem to be as follows. If the output of sulphur dioxide (and, probably, of oxides of nitrogen, though their effects have not been determined) is reduced, there should be a general improvement in all the freshwater concerned. We do not know how great the reduction in output will have to be to restore conditions fully to those which obtained prior to acidification, nor how rapid any improvement will be. In the meantime, the addition of lime or limestone can have a beneficial effect, and where the application can be made to the catchment area, relatively long-term results should be obtained.

Chapter 8

PESTICIDES

There is considerable public concern about the way in which toxic pesticides, particularly those used in agriculture, may get into the environment. We have many alarming reports of their presence in drinking water and food, and of their effects on wild birds and other wildlife.

At first I had some doubts as to whether this was a suitable topic to be discussed in this book. Pesticides – and here I mean insecticides, herbicides and fungicides in particular – are poisonous substances used to kill organisms which we wish to eliminate because they damage our crops and reduce their yield. If all the pesticide reached its target and none was 'lost', there would not be a problem of wastes. But, unfortunately, this sort of efficiency seldom exists. Where an insecticide is sprayed to control a comparatively rare insect pest which is present in small numbers, less than one per cent of the spray reaches the target and over 99 per cent is wasted, and this waste material may have important ecological effects. Not all pesticide applications are anything like as inefficient. Thus when the herbicide paraquat is sprayed on grass which the farmer wishes to kill, very little falls anywhere except on the grass where it is rapidly absorbed and where its biological activity is eliminated. These are the two extremes – with most pesticide applications there is a substantial waste of the chemical, and for this reason the subject will be treated here.

The subject is one of considerable magnitude. Table 10 (p. 116) shows the areas of farmland in England and Wales treated with the different pesticides. Almost all arable land, and even some grassland, is being treated. This costs the farmers a considerable amount, for in 1987 they spent some £195,000,000 on herbicides, £31,000,000 on insecticides and £99,000,000 on fungicides (the British pesticide industry also sold £563,000,000 worth of their products abroad).

There are those who consider that the use of pesticides is unnecessary and could be stopped without seriously damaging our food production and that the use of pesticides has often exacerbated the problem. I consider that this is nonsense. Worldwide it is calculated that nearly 40 per cent of all food produced is lost to pests of some sort. While it is true that some pest problems have been exacerbated by the unwise use of certain chemicals, it is not true that this damage is a new problem. Thus in the Bible there are many references to the devastating effects of insects and of weeds. To quote a few: in Psalm 105, we read 'He spake and the locusts came, and caterpillars, and that without number. And did eat all the herbs in their land and the fruit of their ground.' In the book of Job, chapter 36, it says, 'Let thistle grow instead of wheat and cockle instead of barley' (another version puts 'noisesome weeds' instead of cockle). There are many other examples. Several of the plagues of Egypt described in Exodus were caused by insects, and the importance of weeds is stressed in several of the New Testament parables. The Israelites would certainly not have agreed that they faced no problems from pests in their agriculture.

Similar stories can be found throughout the world's literature. Today in many tropical countries we see whole crops being destroyed when pesticides are not available. In Britain such total destruction is seldom experienced, but pesticides have greatly increased yields, and have made farming more flexible so that farmers have a greater choice of crops to grow at any time, so they are better able to meet the needs of the customer as well as the possibility of obtaining better returns for their efforts.

It is unfortunate that people often talk of 'pesticides' as if they were all very similar chemicals with similar properties, including toxicity. In fact many of the most widely used herbicides are almost non-poisonous to man and other vertebrates, while some insecticides can cause serious damage if ingested in quite small amounts. The situation is well illustrated by the way the European Community has proposed draft regulations for the levels of pesticides which may be allowed in drinking water throughout the community. These regulations suggest that no pesticide should be permitted at a level above one microgram per litre of water, and that the total level of all pesticides should not exceed five micrograms per litre. One microgram per litre is equivalent to one part in a thousand million (or one part in an American billion or in a European milliard). There has been a considerable outcry from some environmentalist groups in Britain, as some samples of our water have contained pesticides in amounts above the levels permitted by the draft directives. They have castigated the British water authorities, and have suggested that consumers are in danger of being poisoned.

In fact no pesticide has been found in any drinking water at a level which his anything like that liable to cause any harm. The most commonly found pesticides (which have only been found in a tiny minority of cases) are the herbicides atrazine and simazine, persistent herbicides used on a variety of crops. These, and other pesticides, have been extensively tested by the Fish and Wildlife Service of the United States Department of the Interior. They found that they obtained no mortality at all when four species of birds were exposed to high doses of these herbicides, including water containing 5,000 parts per million. This harmless concentration is 5,000,000 times the maximum level permitted by the EC draft regulation. For most foods, a rule of thumb is to ban levels of toxic substances more than one fiftieth of that which causes damage. It is certainly time that the EEC should think again, and produce recommendations relating substances to their actual toxic properties.

The trouble is that most people are unable to distinguish between the data relating to high and low levels of pesticide contamination. If a chemical analysis shows that some pesticide can be detected at all, they assume that this is in danger, whereas if the level is very low it may be a reassurance. I was recently asked by the Royal Society of Chemistry to address one of their conferences on the subject: 'The analytical chemist is the cause of most pollution'. This title was not of course meant to be taken literally, but was worded so as to draw attention to what is a very real problem. Twenty years ago, when there was concern about the insecticide DDT getting into foodstuffs, in the USA there was a regulation allowing only 'zero tolerance' of this chemical in any sample of food offered for sale. This was quite sensible, for the analytical methods then in use could not detect DDT at levels as low as one part per million, a level not so far below that which could be toxic. However, today our analytical techniques are as much as a million times more accurate, and these can detect infinitesimally small amounts of DDT almost everywhere, at levels very much lower than those which might cause damage.

Some people may take a different attitude, and say that we cannot be absolutely sure that even the lowest concentration of a poison cannot do some harm. What they should realise is that almost every foodstuff, no matter how carefully produced, naturally contains far higher levels of potentially toxic substances, for instance of metals such as lead and mercury, than it does of insecticides or similar chemicals. In vegetables including potatoes poisons like solanin occur at levels which could harm someone eating ten kilograms or more in a day; to obtain the same harmful effect from pesticide residues would usually necessitate consuming an impossible ration ten or more times as great.

In fact in certain cases low levels of pesticides in animals may actually be beneficial. There is a quite common phenomenon known as 'hormesis' which describes the way in which substances harmful at high concentrations may have the opposite effect at low. This is true of many poisons, some of which are used as drugs in small quantities. In many experiments low levels of DDT have been shown to have beneficial effects on metabolism. As the dose increases we first reach the threshold value where, in these cases, no effect can be detected. Only above the threshold can any damage be detected. The general conclusion is that only relatively high levels of most pesticides do any damage to man and other vertebrates. In this chapter we are therefore concerned with the ways in which waste pesticides can occur at levels where their effects can be demonstrated. These effects, as will be shown, may sometimes arise even when particular substances are properly and correctly used. However, the most serious environmental effect usually only arises from some accident, when high levels of toxic substances are accidentally introduced into the environment.

Herbicides

As will be seen from Table 10 (p. 116), in England and Wales some eight thousand tonnes of the 'active ingredient' of herbicides is used every year. This is nearly eight times as much as the quantity of all other pesticides – insecticides, fungicides etc – which are used. Although, as indicated above, some herbicides such as paraquat produce little waste material to enter the environment, other chemicals are not so efficiently applied. Nevertheless herbicides have comparatively little harmful effect on the countryside or on its wildlife, except of course they do kill many wild plants – weeds – which is the reason for their application.

Before the 1939-45 war, most of the weedkillers available were highly toxic substances, like sodium arsenite or copper salts. These left persistent residues in the soil, and eliminated earthworms and soil fungi. Fortunately they were only used over comparatively small areas, and so they did little harm to wildlife, though they could kill those organisms with which they came in contact. The most commonly used 'total weedkiller' was sodium chlorate, which was applied to paths and other areas where it was intended to eliminate any plants. This chemical was effective for many months. It was washed out by rain, and got into watercourses where, unless considerably diluted, it continued to kill the plants. It had the disadvantage that under dry conditions it readily exploded, and could cause serious fires when used around buildings.

With such a poor collection of weedkillers at their disposal, farmers relied mainly on cultivation and hand hoeing where practicable. Improved seed cleaning eliminated the infestation of cereal fields by such weeds as the corn cockle, *Agrostemma githago*, something often deplored by conservationists who appreciated the beauty of this species, even though it was poisonous. However, a range of small weeds was generally tolerated in most crops, when little harm was done and some benefit was obtained as they improved the conditions for partridges and other birds.

In 1942, an entirely new type of herbicide, the phenoxyacetic acids, or 'hormone weedkillers' started to be used, and are the main methods of controlling weeds in cereal crops today. Probably the substance most widely used is MCPA (4-chloro-2-methyl-phenoxyacetic acid) which is not at all toxic to insects and vertebrates. This is sometimes a disadvantage, as insect pests such as the fruit fly, *Oscinella frit*, appear to survive unharmed when infested oats are sprayed with MCPA to eliminate broadleaved weeds. Tests by the US Fish and Wildlife Service found that concentrations of MCPA of 5,000 parts per million had no harmful effect on quail, pheasant or mallard, and tests on other vertebrates gave similar results.

Under favourable conditions for spraying, that is when there is little wind, most of the MCPA falls on the crop, its weeds or on bare soil. It is not very persistent, and in soil is broken down by bacteria. However, spraying does take place when the wind is blowing, and so quite a proportion is 'wasted' and lands outside the crop. There are many complaints when this happens, although the damage is seldom serious. Herbaceous plants and even shrubs may show the typical distortions these hormone weedkillers produce, and some plant deaths may take place, but the spray becomes more dilute the further it strays from its target, and damage is generally slight and transitory.

Total weedkillers like paraquat and diquat are very poisonous, and have caused deaths when they have been accidentally drunk by children when they have been stored in soft drinks bottles (generally when their parents have 'won' small amounts from their farmer employers to use on their gardens). When applied to plants or to the soil, these chemicals are normally absorbed rapidly and so they do not cause any damage except if spraying is careless and the wrong area is sprayed. Recently glyphosate ('Roundup') has been increasingly used. This is particularly lethal to difficult weeds like couch grass and bindweed. It has the advantage that it is a systemic herbicide, that is if part of a plant is wetted with it, the chemical is translocated to other parts of the plant including underground rhyzomes. Glyphosate is rather expensive, so little is wasted and it has few environmental effects. It is also of very low toxicity to vertebrates.

The long-lasting herbicides simazine and atrazine have been mentioned above. They are widely used on soils on paths and railway tracks where it is hoped to prevent any plants from growing for several months after each application. The chemicals are for the most part bound to the soil, and there is no serious waste problem caused by their being leached out. Some losses do occur, and they have been detected in drinking water (see p. 111) but at such low levels as to be quite harmless.

Herbicides are useful mainly because they make the practices of ordinary farming much easier. Weed control which otherwise would take many hours of labour can be quickly achieved in a short time by spraying the appropriate compound. As a rule the side effects of using herbicides are not serious. Environmentally the main complaint is that they make possible substantial environmental changes which might otherwise be difficult to achieve. There are some effects from the materials which are wasted when they do not reach their targets. More serious damage has been seen in ditches and watercourses where containers not quite empty have been dumped. There have been accidents where serious spillages into similar places have occurred, but generally, when properly used, the harmful effects have been small. Modern herbicides, though used in great quantities, are mostly non-toxic as compared with the substances used even twenty years ago.

Fungicides

Fungal diseases of crops can be very serious. The infamous potato blight reduced the population of Ireland by nearly 50 per cent in the middle of the nineteenth century. This was caused by the fungus *Phytophthora infestans*. This parasite is still common, but it can be controlled by several modern fungicides, so serious crop losses are generally avoided. Cereals suffer from seed- and soil-borne fungal diseases, from mildews and rusts, all of which can greatly reduce yields. Fruit trees are subject to scab, mildew and canker, all caused by fungi. There is therefore a real need for efficient fungicides.

In the nineteenth century Bordeaux mixture, prepared by mixing copper sulphate, lime and water, was used to protect vines in Europe from fungal attack. It and other preparations based on the fungicidal properties of copper were used on potatoes and other crops. When copper was used year after year as in apple orchards, it remained in the soil in concentrations which eliminate most of the soil fauna including earthworms. Sulphur in various forms, as dusts and washes, was widely used. This seems to have had few detrimental effects except to the target fungi.

Most cereal crops are treated with seed dressings to prevent such diseases as bunt, which is caused by the fungus *Tilletia caries*, which turns the grain into a revolting black ball of spores. This has been almost completely eliminated by mercurial seed dressings. As the mercury in the seed dressing is left in the soil, it qualifies as a waste product, and might be expected to have some harmful side effect. However, the amount used is so small as to be harmless – only about one milligram is added to every square metre of the field in which the seed dressing is used. This is a small fraction of the amount of mercury present naturally in most soils.

These seed dressings have killed birds which have dug up the seed and eaten it. Various organomercurial salts, with different vertebrate toxicity, have been used. In Britain, phenylmercury compounds, of low toxicity, have normally been used. In Sweden, where much more toxic methylmercury compounds were employed, bird deaths have occurred. Predatory hawks have also been affected when they ate seed eating birds which had fed on treated grain. However, the situation has been complicated in Scandinavia, where large amounts of mercury have been used as 'slimicides' by the timber industry. This has adversely affected life in fresh water.

Most cereals are regularly treated with systemic fungicides. Unfortunately some of the disease-causing fungi have developed resistance to the fungicides and farmers have had to ring the changes to find a new chemical which is still effective.

Although fungicides are so widely used, they have comparatively little effect on the environment outside the target area. This is because they are generally used in small doses, and also the modern organic compounds like Benomyl have a very low toxicity to both vertebrate and invertebrate animals, and they have little or not effect on growing plants.

Insecticides

Although, as Table 10 shows, we in Britain use comparatively small amounts of insecticides, public concern about these substances is greatest. In fact for many people 'pesticides' is a synonym for 'insecticides'. Britain is fortunate in that it is rare for any crop to be totally destroyed by insect pests, something which commonly occurs in the tropics. If we had no insecticides farming could go on much as at present. There would be serious losses in cereals, beans, vegetables and fruits, and the cosmetic quality of fruits and vegetables would be reduced (as it is with some of the 'organic' produce I see on sale in Cambridge) but our population would not starve.

Before 1939 our supply of effective insecticides was limited. Some were very poisonous, including hydrocyanic gas used as a fumigant, and copper arsenite ('Paris green') for application against caterpillars and

116 WASTE AND POLLUTION

Table 10 Tonnes of active ingredients of various pesticides used on cereals and on other arable crops in England and Wales. Note that the areas treated refer to treatments. Some were sprayed more than once. Almost all the 3,600,000 ha of cereals were sown with dressed seed (3,358,000 ha), an operation which can only be done once! Some had no herbicide treatment, some had more than one spray, hence the total of 4,408,000

	Hectares treated	Tonnes of active ingredient
Cereals		
Organochlorine compounds	1,000	6
Organophosphorus compounds	294,000	107
Other insecticides	272,000	43
Seed treatments	3,358,000	48
Fungicides	978,000	588
Herbicides	4,408,000	8,026
Other pesticides	188,000	263
Other Arable Crops		
Insecticides		
Organochlorine compounds	40,000	35
Organophosphorus compounds	274,000	99
Other insecticides	180,000	416
Seed treatments	430,000	2
Fungicides	616,000	882
Herbicides	918,000	6,131

beetles. Sodium fluoride was used against ants invading houses, and nicotine to kill aphids in glasshouses. The safest preparations were made from natural products, rotenone from *Derris elliptica*, and pyrethrum from the flowers of *Chrysanthemum cinerariaefolium*. These vegetable products had the advantage that they were comparatively non-poisonous to mammals and birds, though derris was used as a fish poison, which did not render the catch dangerous to man. These vegetable insecticides break down quickly, especially if exposed to light, so they never became long-term environmental pollutants. In view of the concern at the persistence of new insecticides in the post-war period, it is somewhat ironical that research was going on to try to prolong the activity of pyrethrum. The disadvantage of the vegetable insecticides, which became very real during the 1939-45 war when there was a risk that insect-borne diseases like typhus and malaria might cause more casualties than enemy action, was that they could only be produced in limited amounts, and that to increase the crop would have taken years to organise.

During the 1939-45 war two groups of new insecticides came into use. The first was the organochlorines, or chlorinated hydrocarbons, of which DDT is the most familiar. The second group, the organophosphorus compounds, was discovered in work with allied compounds like mustard gas which were being tested for use in chemical warfare. Parathion (a deadly poison to man) was the first member of this group to be widely used. The great virtue of these synthetic chemicals was that industry could produce them in unlimited quantities.

DDT was synthesised in the laboratory in 1874, but its insecticidal properties were only discovered by Dr Paul Mller of the Swiss Geigy company in 1939; a patent application was made in Switzerland on 7 March 1940. It was first widely used to control the Colorado beetle, which threatened to destroy the Swiss potato crop, in 1941. Samples became available to the British and American authorities, and the chemical was manufactured in considerable quantities in those countries. It was widely used during the latter days of the war, when its use saved many millions of lives. In 1943 when the Allies occupied Naples they found that a typhus epidemic, the disease being carried by the body louse *Pediculus humanus*, had begun, and was causing serious civilian casualties. Had the conditions which existed in the war of 1914-18 been present, the epidemic would have spread with horrendous casualties among both civilians and the military. As it was over two and a half million individuals were rapidly treated by having dust containing DDT liberally blown into their underclothing. This was particularly effective in the least hygienic members of the population who seldom changed their garments, but everyone appeared to be protected. The cases decreased and stopped entirely, and the military appeared to be completely protected. In other theatres of the war, DDT was extensively used to control malaria-carrying mosquitoes.

The most notable feature of the insecticide DDT was its lack of toxicity to man. No serious side effects were noticed in Naples even among those most heavily dosed. In fact DDT has been found to have an acute toxicity similar to aspirin. No deaths from its use have ever been reported, except where it has been eaten in large quantities having been mistaken for flour. At this date it seemed to be the perfect insecticide for use against a wide range of medical and agricultural pests. However, there were those who sounded a word of caution. The earliest published warning which I have discovered was one which I gave at a meeting of the Royal Society of Tropical Medicine in London on 15 February 1945. I said:

"DDT was probably the greatest advance in insect control which had ever been made. But DDT was clearly no panacea, which could be broadcast indiscriminately to kill all noxious pests. A great increase in field research was necessary wherever DDT was used. Fortunately mosquitoes and muscid flies seemed particularly susceptible to this substance, but all arthropods were affected to a lesser or greater extent.

Much work should be done on its effects, in the field, on all manner of apparently unimportant insects and other forms of life, to ensure that there was not a serious upset of the 'balance of nature' with subsequent disastrous effects."

I went on to point out that my own experiments had shown that it was less effective against Arachnida, including the scabies mite, *Sarcoptes scabiei*.

By the mid-1950s there was more widespread concern about the ecological effects of DDT and the other chlorinated hydrocarbons. In Britain large numbers of seed-eating birds were found dead in the fields in spring, and there were disturbing reports of a fall in numbers of breeding peregrines and sparrowhawks, possible damage to golden eagles and buzzards, and deaths of foxes and badgers. Eventually this was found to be caused by seed corn dressed with aldrin or dieldrin to protect the young seedlings from attack by the larva of the wheat bulbfly, *Leptophylemyia coarctata*. What is ironical is that the damage was done as a result of trying to reduce the risk of widespread pollution resulting from a considerable waste of the insecticides which would have occurred had they been broadcast among the crop. It was decided to stick a minute amount of the insecticide onto each grain, so as to concentrate it at just the spot where the pest attacked. This reduced the total amount of dieldrin or aldrin applied by a factor of ten, but unfortunately the reduced amount of poison was in just the situation where it was ingested by seed-eating birds, who then passed it on to their predators.

There is a comparatively happy ending to this particular story. On the recommendation of the Advisory Committee on Pesticides, it was agreed that seed dressed with aldrin and dieldrin would no longer be used in spring-sown corn. This was a voluntary but highly effective arrangement, which virtually stopped the deaths of seed-eating birds in subsequent years. Incidentally this effective action in Britain was taken in June 1961, some time before the publication of Rachel Carson's *Silent Spring*, and it shows that scientists were well aware of the dangers of pesticides long before this book appeared.

In Britain massive deaths from pesticides, such as those which occurred from these seed dressings, have otherwise been uncommon. I know of no cases where DDT has acted as an acute poison, but it has clearly had important ecological effects, and these have resulted in it, and the chlorinated hydrocarbons, being virtually withdrawn from use. Although the acute poisons ceased to be widely used in 1962, DDT was still generally available, notwithstanding increasing ecological evidence that some species of wildlife were having their populations seriously reduced. At first this was difficult to understand. It was found that predatory birds were indeed picking up DDT, but levels in their tissues were generally well below those known to be harmful. This was largely because the DDT was, as we thought, 'detoxified' by being transformed by metabolism to DDE, a compound of much lower toxicity. However, eventually we found that DDE, particularly in birds, has other properties. It is responsible for causing the birds to lay eggs with thin shells which are easily broken in the nest and which reduce breeding success.

DDT was also found to get into freshwater. It was very insoluble, and levels were too low to affect animals drinking the liquid. However fish, which 'breathe' very large amounts of water to obtain their oxygen, also took the insecticide into their bodies via their gills, and concentrated it by a factor of as much as ten thousandfold. Its use, despite its other values, has therefore been greatly reduced throughout the world, and it has been banned in most developed countries. Its use has continued in countries which cannot afford more costly and less effective chemicals, and where insect-borne diseases are causing many deaths, though even here the search for substitutes continues.

DDT is responsible for making an unimportant pest more serious. As mentioned in my statement to the Royal Society for Tropical Medicine, Arachnida are less seriously affected than insects. Codling moth and other pests of fruit trees have been effectively treated with DDT, but in many cases this has resulted in a great increase in the level of Red Spider Mite, *Tetranychus telarius*, which has proved more difficult to control.

Because of its persistence, many people expressed the fear that even if DDT ceased to be used, its levels in the environment would continue to rise and give serious results for many years. Thus in *Limits To Growth* we read that DDT levels in fish will continue to rise for twenty or so years even after its use stops. Fortunately this has not occurred. Since the 1970s DDT levels in birds and in the environment have decreased considerably. Today the predatory birds usually lay eggs with normal shells, and populations of such species as the Peregrine are back to pre-1939 levels – with the result that racing pigeon owners are once more demanding that their numbers should be controlled!

The first organophosphorus insecticides, parathion and TEPP, were acute poisons which killed many operatives. Unfortunately when DDT was prematurely banned in some tropical countries, parathion was reintroduced and many humans died. These substances are quickly broken down, so do not act as long lasting pollutants. Fortunately less toxic products such as malathion are now widely used, with little damage to operatives or the environment.

The general trend in recent years is for the use of less poisonous pesticides, and for substances of only limited persistence. The public is much less at risk than it has been in the past. Even before 1939 there were cases of poisoning with arsenic which has been used on apple trees. It is somewhat ironical that there is probably more public concern about the danger of almost non-existent pesticide residues in our food than there was when this was a real problem.

Chapter 9

WASTE HEAT

All energy ends up as waste heat. Our own metabolism, which keeps our bodies warm, also produces heat which escapes into the surroundings. In temperate or cold countries this waste heat produces a local warm micro-climate which supports its own fauna and flora. Thus the body louse lives in human clothing and feeds by sucking blood. But it is equally dependent on a warmer environment, for this insect could not normally survive or breed at the sort of outside temperatures usually found in Britain. In fact, people who live in very cold countries, with temperatures well below zero, produce body heat and support parasitic insects and other forms of life on their skin or among their garments which would not otherwise survive.

Metabolic heat, however, is only a fraction of what we liberate by burning fuels and by exploiting the various sources of energy on which civilisation depends. There have been worries that the greatly increased amount of heat might be enough to affect the whole globe, with unpredictable results. This widespread effect is unlikely, for man's present emissions of waste heat, though very substantial, amount only to about 0.01 per cent of the heat received from the sun. Even if the standard of living everywhere was raised to that of the current level enjoyed in the USA, where energy is spent and wasted prodigally, the total heat given out would be less than 0.1 per cent of the sun's contribution.

Nevertheless, the heat we produce can have important local effects. Our cities are significantly warmer than the surrounding countryside; the mean annual figure for this warming in London is about 1°C (Fig. 22). In calm, cold weather in winter, when heating of buildings is at a maximum, the warming is greater and may even rise to about 11°C.

In New York the heat emitted is said to exceed that received from the sun by a factor of 7. The temperature rise there may be greater in summer than in winter, caused by the excessive use of air conditioning, which cools the insides of the buildings, but which transfers this heat, plus what is released by the inefficient use of electricity, to the outside. One other cause of high temperatures, particularly at night, in cities is

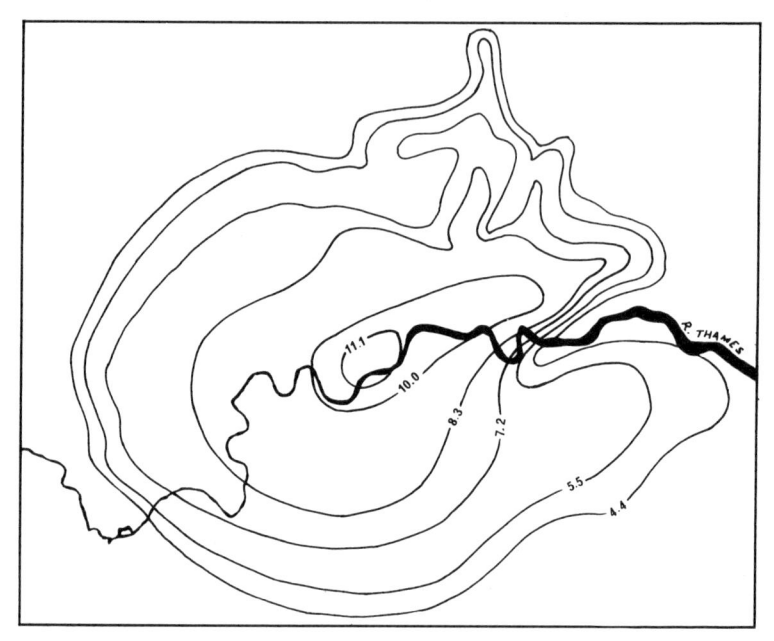

Fig. 22 Minimum temperature distribution in London on 14 May 1959 (After Chandler 1965)

that the buildings absorb heat from the sun during the day, and this is radiated out during the night.

Warm city centres may support populations of plants and animals which are intolerant of cold. Thus frost is less common in the city, so that species which would be killed elsewhere survive at least the earlier part of the winter in urban gardens. These surviving plants may provide food for animals. The massive numbers of Starlings which roost in Central London, and which make such a mess of buildings, are probably the result of the warmer city climate. Feral Pigeons throng our cities largely because of the food available, much of it provided by generous visitors in defiance of requests by the authorities not to feed the birds. The warmth and the extra light (partly a form of waste energy) enables Pigeons and Sparrows to extend their breeding season; also they need less food to keep warm, and so more survive than would under colder conditions.

The waste heat so far considered enters the environment more of less accidentally. The greatest local effects arise when industrial plant, particularly electricity generating installations, are designed so as to use water to cool their machinery. Here the effect is mainly restricted to the water into which any heated effluent is discharged, but the effects on aquatic forms of life may be very considerable. Small power stations have drawn cooling water from streams, passed it through their machinery, and released the warmed water to the stream. They have also used stream water in the operation of their boilers. However, large power stations such as those operated in Britain by the Central Electricity Generating Board need such quantities of water that they would dry up most available natural sources; thus most cooling water is recirculated time and time again, and only comparatively small amounts are returned to affect the river, lake or the sea from which it was originally derived. In these cases the greatest part of the heat loss is by evaporation in cooling towers, but the not inconsiderable residue goes to raise the temperature of the receiving water.

One important effect of raising the temperature of water is that this also reduces its capacity to dissolve oxygen, as was previously mentioned in Chapter 7. Thus a litre of water, at 5°C, in free contact with the atmosphere, can absorb about 9 ml of oxygen weighing 13 mgs. As the temperature rises, the oxygen content falls, so that at 20°C it is only about two-thirds of the level at 5°C. As the rate of metabolism of cold-blooded animals, over the range of temperatures at which they normally live, may double with a rise of temperature of 10°C, one would expect an oxygen shortage. It should be noted that air is much richer than water as a source of oxygen. A litre of air contains 210 ml of oxygen, weighing 300 mgs, i.e. over 20 times as much as is found in the same volume of well-oxygenated water. So a fish has to pass a great deal of water over its gills to breathe, and the lower the oxygen content the greater the amount of water. As the gills absorb toxic substances as well as oxygen, this may explain why these are particularly dangerous in polluted and thus often partially deoxygenated water.

The question of deoxygenation by sewage and other pollutants is dealt with in Chapter 7 (p. 123). Cases of deoxygenation by heat do occur, but are relatively uncommon. In fact, warm water leaving an electric power station may contain more oxygen than was found in the water, at a lower temperature, when it was abstracted from the river.

At Drakelow on the River Trent, for example, some 850,000 m^3 of effluent from the cooling towers was returned to the river every 24 hours in summer, and 650,000 m^3 in winter. Less water was required in winter as it started off colder, so that it could absorb more heat. The average temperature of the intake in summer was 18.2°C and that of the outlet was 27.4°C. In winter, the corresponding figures were 6.6°C and 20.4°C. Rises of this magnitude obviously had a substantial local effect, and temperature rises, to 22.4°C in summer, and to 12.4°C in winter, were measured 1.6 km downstream; so a large area of the river, and a large volume of its water, was affected.

It is perhaps surprising that this warming of the effluent did not cause any deoxygenation. In fact, the oxygen level in this river was substantially improved. The Trent is a moderately polluted river, with the oxygen level falling as low as 17 per cent of saturation, and seldom rising above 50 per cent. The water, as it passes through the cooling towers, becomes saturated, or even, to some extent, super-saturated. As it became warmer, the amount of oxygen taken up was limited, but nevertheless at Drakelow between 4 million tonnes of oxygen in summer, and 2 million tonnes in winter, are added to the river every day. Of course, had the river been more highly oxygenated from the start, a different result would have been experienced. But as a rule, heating from cooling towers seldom produces any problem from deoxygenation; as will be seen, high temperatures and toxic biocides are more likely to be harmful.

As mentioned above, waste heat mainly affects the environment by warming water used to cool industrial plant, especially electric generators. Aquatic organisms, other than warm-blooded mammals and birds, have a similar pattern of reactions to temperature. They have a cold death point, where exposure is lethal. Above that is a chill coma temperature, where they are immobilised, but generally survive this state, perhaps for very long periods. At the upper end of the scale is a heat coma temperature, and above that a thermal death point. As a rule, the heat coma temperature is only a degree or two below the thermal death point. The thermal death point may not be sharply defined – organisms may survive for short periods at temperatures which, with longer exposures, are lethal. In between the chill coma temperature and the organisms' upper limiting temperature there is usually a narrower zone where optimum conditions prevail.

The position of these various points on the temperature scale may vary greatly for different organisms. As a general rule, those from warm environments are more liable to damage, and to immobilisation, at temperatures where organisms from cold climates are able to live normally. Thus warm effluents, which may be unsuitable for many native species in Britain, are sometimes invaded by aliens from hotter countries.

In natural freshwaters in Britain, the assemblage of plants and animals in a cold, upland stream differs from that of a warm river at a lower level in the south of England. Many animals found in cold waters cannot survive in warmer waters which provide optimum conditions for other species. Thus the flatworm (*Planaria alpina*), which is restricted to cool springs, dies when subjected to temperatures above 28°C, which are both civilians and the military. As it was over two and a half million individuals were rapidly treated by having dust containing DDT liberally blown into their underclothing. This was particularly effective in the least hygienic members of the population who seldom changed their garments, but everyone appeared to be protected. The cases decreased and stopped entirely, and the military appeared to be completely protected. In other theatres of the war, DDT was extensively used to control malaria-carrying mosquitoes.

The most notable feature of the insecticide DDT was its lack of toxicity to man. No serious side effects were noticed in Naples even among those most heavily dosed. In fact DDT has been found to have an acute toxicity similar to aspirin. No deaths from its use have ever been reported, except where it has been eaten in large quantities having been mistaken for flour. At this date it seemed to be the perfect insecticide for use against a wide range of medical and agricultural pests. However, there were those who sounded a word of caution. The earliest published warning which I have discovered was one which I gave at a meeting of the Royal Society of Tropical Medicine in London on 15 February 1945. I said:

"DDT was probably the greatest advance in insect control which had ever been made. But DDT was clearly no panacea, which could be broadcast indiscriminately to kill all noxious pests. A great increase in field research was necessary wherever DDT was used. Fortunately mosquitoes and muscid flies seemed particularly susceptible to this substance, but all arthropods were affected to a lesser or greater extent. Much work should be done on its effects, in the field, on all manner of apparently unimportant insects and other forms of life, to ensure that there was not a serious upset of the 'balance of nature' with subsequent disastrous effects."

I went on to point out that my own experiments had shown that it was less effective against Arachnida, including the scabies mite, *Sarcoptes scabiei*.

By the mid-1950s there was more widespread concern about the ecological effects of DDT and the other chlorinated hydrocarbons. In Britain large numbers of seed-eating birds were found dead in the fields in spring, and there were disturbing reports of a fall in numbers of breeding peregrines and sparrowhawks, possible damage to golden eagles and buzzards, and deaths of foxes and badgers. Eventually this was found to be caused by seed corn dressed with aldrin or dieldrin to protect the young seedlings from attack by the larva of the wheat bulbfly, *Leptophylemyia coarctata*. What is ironical is that the damage was done as a result of trying to reduce the risk of widespread pollution resulting from up the circulation. These biocides have proved very effective in reducing the invasion of mussels in cooling systems that use sea water, and the raised temperature has probably increased their toxic effect.

When an electric generating station is shut down in winter, the temperature falls substantially. In the USA this has been found to kill fish. Apparently, species which would otherwise migrate downstream in autumn, to avoid the cold, have remained in the favourable conditions of the thermal plume, and have then been subjected to this sudden fall in temperature.

Where there is a local rise in temperature to a level which might be harmful, killing sedentary organisms, fish and rapidly moving invertebrates may escape to cooler waters and thus avoid damage. Heat, by increasing the rate of metabolism, will make the fish move more swiftly and thus they are more likely to escape danger.

In Britain, where waters have rarely been heated to dangerous levels, there have been three main effects. First, there has been a change in the composition of the population. Secondly, there has been an extension of the breeding season of organisms which are inactive in cold conditions. Thirdly, there has been an invasion by exotic species which require warm conditions. Many examples of these changes have been given in two volumes on the effects of thermal changes arising from electricity generation by Dr T.E. Langford (1983, 1990) and only a few characteristic cases will be quoted here.

Whereas warm waters cause a change in the numbers of different species, as compared with unheated reaches, the variations have generally been less than might be expected. Stonefly larvae (*Plecoptera*), which are particularly susceptible to pollution as well as to raised temperatures, are often eliminated, and the general effect has been a reduction in diversity.

Warming water is apparently beneficial in that it encourages breeding and also extends the period when growth takes place. This has been exploited by those concerned with Oyster and Mussel culture. Many invertebrates evidently continue to breed in autumn in rivers such as the Severn below the Ironbridge power station, when the same species above the power station are more or less dormant.

Important invasions of exotic species in heated British rivers have occasionally been reported. In reaches of the Trent, the exotic Mediterranean snail *Physa acuta* has tended to replace the less heat-tolerant indigenous *P. fontinalis*, and is quite widespread, having been found in more than half the power station effluents studied. The Asian worm *Branchiura sowerbyi* occurs in Britain, but is less common than in America. Considerable numbers of immigrant warm-water marine species have been found in heated waters in the Swansea docks, and the populations appear to be reinforced by specimens brought in by shipping. The exotic barnacle *Balanus amphitrite* and the wood-boring isopod *Limnoria tripunctata* have replaced indigenous species. Damage done by *Limnoria* and also by the shipworm *Teredo navalis* has been exacerbated by the warm conditions which have allowed these animals to extend the period of their activity into the winter. There are few reports of exotic species in rivers, doubtless invasions seldom occur in inland waters.

The comparative unimportance of heat as a pollutant in Britain, and even in America (where many more incidents have been observed), may come as something of a surprise. The position is well summed up by T.E. Langford in the concluding section of his recent book: 'It seems clear that the prediction of the effects of thermal discharges from the new breed of power stations need not be subject to the irresponsible extrapolation and exaggeration which was found in the 1960s and early 1970s. The research and surveys have not borne out the dire predictions of disaster which came from academic and political ambition.'

It is perhaps surprising, in these days when there is so much talk (if so little action) about 'recycling', that more use is not made of all the waste heat generation in our power stations. The trouble is that such waste is so-called 'low grade heat', where temperatures are not very high, so that its use is limited. Most of our power stations, too, are some distance from centres of population, where the heat could be utilised. In other countries there are successful schemes where combined use is made of heat and power. For these, the whole installation must be properly designed at the outset, and, as already suggested, the power station must be in an urban situation. In fact, one of the first such schemes was operated successfully in London, but was discontinued when the power station was closed and electricity brought from a large generator outside the city. This was justified because the new generator proved much more efficient than the old one, only using half the fuel to generate the same amount of electricity, more than compensating for the saving made in the previous scheme.

The one area in which surplus waste heat has been used successfully is in the heating of glasshouses – there is plenty of room for these in the rural sites where the power stations are now built. Since the heat would otherwise probably be totally wasted, it does not matter that this heating method is rather inefficient. It is likely that this use will increase, and that tropical crops will become a feature of the Trent valley where the main concentration of power stations is found. The other successful use of waste heat is in fish farms, and in raising shellfish, where growth is accelerated by the raised temperature. This is generally uneconomic where the energy has to be paid for at current rates.

Chapter 10

THE GREENHOUSE EFFECT

In a previous book, *Pesticides and Pollution*, first published in 1967, I wrote: 'Recently it has been suggested that CO_2 may eventually have a drastic effect on world climate. Coal and other 'fossil fuels' are being burned at such a rate that the CO_2 content of the whole atmosphere may be raised by as much as 25 per cent by the year AD 2000, and the level will probably continue to rise. The effects of this are not fully understood but some scientists think that the temperature and other properties of the stratosphere may be affected. This could alter the world's radiation balance, possibly melting the polar ice cap. So far little or nothing has been done to reduce the output of CO_2, though some research on ameliorating its possible effects has been suggested. So far most scientists have thought that CO_2 pollution was of little importance; it now seems possible that it may cause greater changes to the world than any other man-made factor in our environment. On the other hand, this may be a completely false alarm.'

The reason many scientists did not think that the rising levels of carbon dioxide would have any important effect on world climate was that they knew that temperatures had fluctuated considerably over the years – there had been Ice Ages, and warmer periods. Since the time of the Norman Conquest, we have had warmer conditions, when vineyards flourished in England, and colder periods. Between 1850 and 1950 our climate warmed by nearly a degree. Sceptics assumed that these fluctuations in temperature were greater than anything that man-made changes in carbon dioxide would be likely to produce.

Today things have changed. There is much wider agreement that the rise in the level of carbon dioxide and the other 'greenhouse gases' will have a serious effect on the world's climate, that it is likely to produce a significant warming, and that the weather pattern in many countries may be changed, so that some parts of the world where important crops are grown to feed people may become too dry, and other countries, now desert, may get enough rain to grow large amounts of food. The warming is also expected to cause a rise in ocean levels, with a risk of flooding in all low-lying countries. What we do not know is when these changes will take place, nor do we know how great the rise in the average world temperature will be. The situation is so complicated, and there are so many conflicting forecasts, there is inevitably support for the view that nothing can be done to alter things, and that the best advice is to 'wait and see'. For reasons which I shall explain, this is not my own view.

I am not sure that it is correct to speak of the 'Greenhouse Effect', as this pre-supposes that something measurable has already occurred. It would perhaps be better to talk of the 'Greenhouse Theory', because all the worries about rises in global temperature and climatic changes are based on the theory that if levels of carbon dioxide (and, equally important, other gases including methane, which are often forgotten by those who make alarmist statements) continue to rise, the atmosphere will still allow the sun's radiation to reach the earth's surface, but the infrared radiation from the earth will be trapped on its way out just as heat is trapped in a greenhouse. It is a fact that higher levels of greenhouse gases do trap infrared radiation in this way; what we do not know is the magnitude of the changes which any particular level of these gases will produce.

The greenhouse effect is clearly a global problem, and we in Britain can play only a minor role in trying to solve it. However, climatic changes, even small changes, will have very important effects on wild plants and animals. A rise or fall of only a couple of degrees could alter the whole nature of our flora and fauna. Species now flourishing could become extinct, and species now excluded from our country could be successfully introduced. Any substantial rise in sea level would also affect the whole shape of our islands, for a rise of quite moderate proportions could flood many low-lying areas and eliminate marshes which are now notable for their wading birds. For that reason the inclusion of this subject seems justified in this book.

First, we must try to separate fact from theory. Levels of carbon dioxide in the atmosphere have certainly been rising for more than 100 years, and they are still rising. We do not have many good records for levels in the last century, but there seems little doubt that, before the Industrial Revolution had exercised its effect worldwide, say about 150 years ago, the level of carbon dioxide was not higher than 275 parts per million (ppm). This figure is confirmed by measurements of the gas trapped in glacier ice (Fig. 23). Today it is approximately 350 ppm. This is a rise of nearly 30 per cent.

The best record of carbon dioxide concentration, from 1958 to 1984, comes from the Mauna Loa Observatory in Hawaii. This site was chosen as it is not near any major sources of emission, and so can be considered

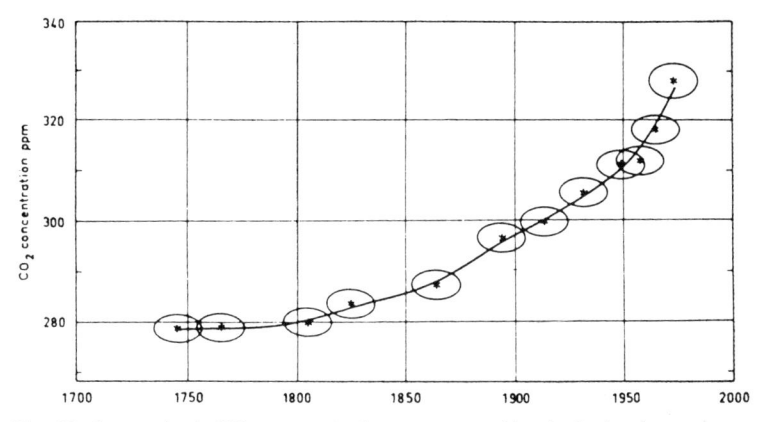

Fig. 23 Atmospheric CO_2 concentrations measured in glacier ice formed during the last 200 years. (After Neftel et al, 1985)

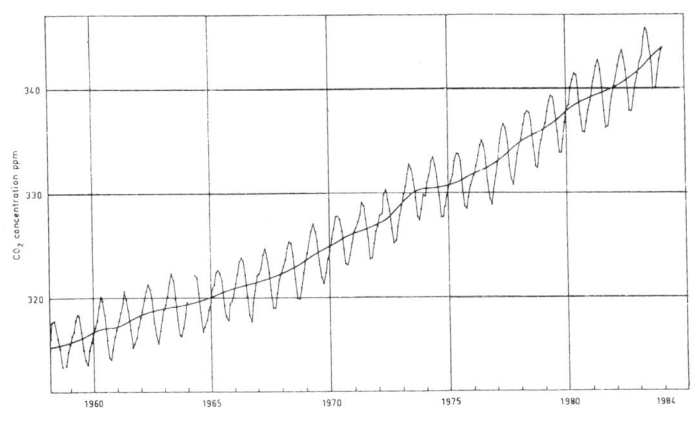

Fig. 24 Concentration of atmospheric CO_2 at Mauna Loa Observatory, Hawaii.

to represent basic global figures for the northern hemisphere (Fig. 24). During the period of 26 years, the mean levels rose from 310 to nearly 350. At the time of writing (1988) the figure must be well above 350. It should also be noted that in each year the level fluctuates considerably, the maximum being some 10 ppm higher than the minimum. The reason for this is that during the summer photosynthesis by crops and forests is much greater than during the colder winter, so greater amounts of carbon dioxide are removed from the atmosphere. This observation is important, and suggests ways in which increased photosynthesis may be manipulated if we wish to curb the rate of increase in carbon dioxide levels.

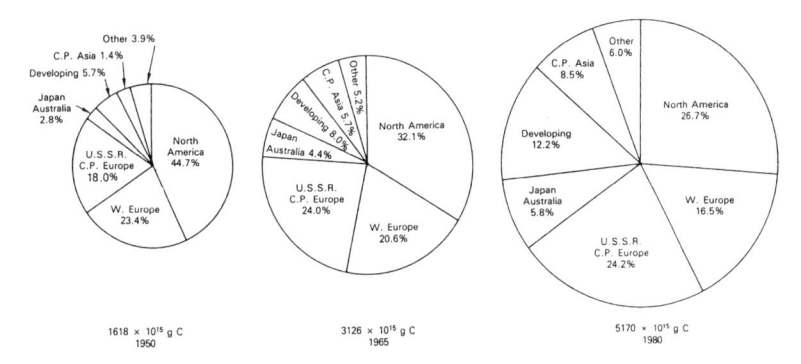

Fig. 25 The changing patterns of global CO_2 emissions, 1950-1980. (Figures from US Dept. of Energy, 1985)

One other point should be noted. Although eventually the atmosphere from the north and south hemispheres becomes mixed, this is a slow process. Nevertheless the land area in the northern hemisphere, particularly where plant growth is vigorous, is much greater than that of the southern hemisphere, so a similar figure derived from southern hemisphere data would probably show much smaller fluctuations.

There is no doubt, therefore, that carbon dioxide levels are rising. What remains unexplained is why they are not rising even more rapidly. The amount of carbon dioxide in the atmosphere represents only about half of that which is being produced, and we are not sure where the rest has gone. The main 'sink' is presumably the ocean, but this gives rise to more uncertainties. Will the oceans continue to take up this proportion of the gas or, as levels rise, will it absorb a greater proportion? On the other hand, if temperatures rise, will less carbon dioxide be dissolved, and instead may more be set free into the atmosphere? We cannot be sure what future atmospheric levels will be, even if we can make a reasonably good guess at the likely range of emissions. Fig. 25 shows how emissions have grown since 1950, and the contributions of different parts of the world. Were the developing countries to give out the same amount as North America and Western Europe, the annual production would be more than doubled, and future extrapolations of the graph in Fig. 24 would have to be shown as rising very much more steeply.

Increasing levels of carbon dioxide will undoubtedly trap more heat, but how much and at what level we do not know. Furthermore, we must remember that there are other greenhouse gases, which could be even more important – for example, chlorofluorocarbons (CFCs), which have been implicated in affecting the level of ozone in the stratosphere, and

THE GREENHOUSE EFFECT   133

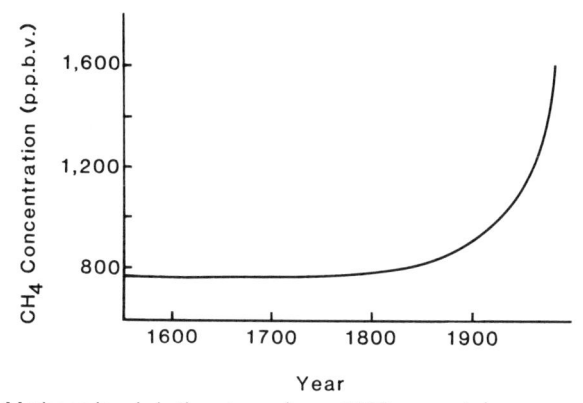

Fig. 26 Methane levels in the atmosphere, 1600-present day

so the penetration of ultraviolet radiation to the earth. This is discussed in Chapter 11. Although CFCs are present in very low concentrations, they are thought to be greenhouse agents some 200 times as potent as carbon dioxide. What is encouraging here is that, because of concerns about ultraviolet radiation, there is international agreement to curb the use of CFCs. Having gone so far, the different countries using CFCs could without any major difficulty or without incurring impossible expenses, accelerate the process and stop emitting any substantial quantities of CFCs within the next few years.

The joker in the pack may well prove to be methane, CH_4, which also has greenhouse properties. Concentrations in the atmosphere have more than doubled since the year 1800, as shown in Fig. 26. This level should, theoretically, have about half the effect of the present level of carbon dioxide. Methane levels are rising fast and may be difficult to control. The figures concerning output of various sources are also very uncertain, the total output being estimated as being anywhere between 275 and 790 tonnes per annum. The main sources are enteric fermentation in livestock and insects, and rice fields and natural wetlands; but biomass burning, landfills and gas- and coalfields also contribute. Since our data for output is still so imprecise, it is difficult to make a reasonable estimate of future effects; and rational control of such sources as ruminants and paddy fields will be very difficult.

A further, relatively small contribution may come from nitrous oxide (discussed on p. 83) and ozone. As there are already reasons for controlling these gases, they need not play a large part in raising global temperatures.

Unfortunately, we have no genuine 'experts' to make reliable predictions as to the probable effect on world climate of these increases in greenhouse gases. What we do have is a tremendous outburst of energy devoted to producing computer-operated models of the possible outcome. These are becoming increasingly sophisticated, and take enormous amounts of computer time, yet the results are almost entirely useless. The best estimates of the global temperature rise by the middle of the next century range from about 1– 5.5°C. Anyone with an inkling of understanding of statistics will realise that with such variations, a fall in temperature could also be predicted! But without reliable data to feed into the computer, such expensive exercises are, in my opinion, a waste of time and money. Even so, the known properties of the gases involved are such that, at some stage, a considerable rise in temperature and an alteration of climatic patterns is almost inevitable.

Why should this be a source of concern? Many people in Britain would welcome a climatic change that resulted in the country having a Mediterranean climate. Moreover, the world and its living organisms have survived great temperature changes in the past, so why should these be so dreaded in the future? Such questions are not difficult to answer. The main problem is that there are now over 5,000,000,000 human beings on the earth, all of whom have the right to the minimum resources necessary for life. When numbers were small, people could move from one part of the world to another. Crops could be grown wherever conditions were suitable. Nevertheless life was often very difficult and there were many disasters, with entire nations devastated by famine and flood. Today it would be impossible to cope with such a change as would occur if the wheat belt in America ceased to be able to feed much of the world. Fairly modest rises in sea level would overwhelm countries like the Netherlands and Bangladesh. Looked at dispassionately, the possible dangers within the next 100 years are such as to make any of our present pollution or other environmental problems seem quite trivial.

It is interesting to note how, in 1990, everyone from the Prime Minister downwards is talking about the greenhouse effect. Some scientists who should know better are citing forecasts of temperature rises as though they were prophesying established facts. One so-called 'expert' from the World Bank is reported as saying that the world has only 'ten years remaining to save itself'. I am prepared to stick my neck out and say that this is a great exaggeration, on a par with the forecasts by some ecologists in the 1960s that the oceans would be dead by 1980. There are those who maintain that the disasters of 1988, for example the American drought and flooding in the Sudan, are indications that the greenhouse effect is already operating. More sober analysts, while discounting these suggestions, are nevertheless talking in terms of when, and not if, such

THE GREENHOUSE EFFECT 135

climatic changes will become significant. So it would obviously be wise to try to devise a global plan of action.

This has already been done in some circles. Various plans for reducing carbon dioxide all have much in common. These were summarised in a letter I wrote to the *Daily Telegraph* on 3 November 1988: 'Increased economy in fuel use, the development of many sources of renewable energy, safeguarding of forests and a massive programme of tree planting.' I also concluded that an enormous increase in the use of nuclear energy must also be included. I shall deal with that later; first I shall consider the other points.

Quite apart from the greenhouse effect, there have long been good reasons for increased economy in fuel use. Fossil fuels, particularly oil, are finite in quantity, and will be exhausted sometime in the 21st century. The rate of depletion is nothing like as rapid as some people foretold 20 years ago, oil prices in recent years have fallen and not risen, but economy is still desirable. In fact, economies have been made, generally by improved efficiency. In Britain today we get twice as much electricity from the same amount of coal as we did 30 years ago; this has reduced carbon dioxide output considerably. There has been a remarkable improvement in the efficiency of the motor car, so that large American limousines use little more 'gas' than did the previous generation of small, compact automobiles. But greater economies, particularly in space heating, are possible, and these need not impair our efficiency or comfort, as has been shown in reports by Gerald Leach and the Watt Committee on Energy. Developed countries could reduce their energy use by 30 per cent, which would leave a little margin for developing countries. No matter how economical we ourselves are, on a world scale the amount of carbon dioxide emitted by different fuels is likely to increase, unless other measures are invoked.

Many environmentalists think that the solution is to shift to much wider use of renewable energy, water power, solar power, wind and tidal energy. This is all very sensible, but can only make a fairly small contribution to world use. It is perhaps ironical that hydroelectric schemes, which do not pollute at all, are generally opposed by the same environmentalists who are advocating the maximum use of clean, renewable energy on the grounds that the dams and other engineering works might adversely affect the countryside. A typical case was in Tasmania, where there is immense scope for hydroelectric development. I think that hydroelectric power should be exploited where possible, and that some environmental sacrifices should be made to that end, but this will only make an important contribution in a few countries. Solar power has a part to play, and this will increase if and when more efficient means of collecting the energy are available. Here again there may be an environmental price to pay. Were it possible to operate solar-powered electrical generating stations, collectors would have to sterilise many square kilometres of the countryside. This would hardly improve its beauty, or its potential to support wildlife.

Wind power should make some contribution. Small wind pumps are used on farms in many parts of the world, and save the expensive installation of electric cables in remote areas. Isolated communities may also utilise wind power. There are many ambitious plans for gigantic 'farms' of enormous windmills. Some of these have been developed in California, where the environment is ruined by huge, noisy construction. There is a place for such installations, particularly in offshore locations, but their total contribution to electric power in any developed country is likely to be small. Tidal power clearly has potentials, too, particularly by damming estuaries (and possibly ruining wild bird sanctuaries). Wave power, exploited by rocking 'ducks' in the open sea, is thought by some to be important, but others are less optimistic. I should be surprised if all the types of renewable energy could generate as much as 20 per cent of that required by any industrialised country.

This brings us to the generation of electricity. As already mentioned, recent improvements in the efficiency of coal-fired stations have already reduced carbon dioxide outputs. If these stations used natural gas, while supplies last, there would be a further halving of carbon dioxide output, due to the properties of the fuel. However, in my opinion, there is no long-term solution except a massive turnover to nuclear power. With ample supplies of electricity we could bring back trams and trolley buses to our towns and cities, and even with present techniques electric cars could carry much of the urban traffic. There must surely soon be improvements which will allow electric cars and buses to cover much longer journeys, so that petrol and diesel vehicles will become minority vehicles. Otherwise we may need to install power lines available to cars on main highways.

Unfortunately there is much public opposition to nuclear power, encouraged by propaganda from environmental groups, and fueled by biased and inaccurate reports from the media. In my opinion, even if some of these reports were true, the risks from nuclear power and nuclear waste would be much less than those which would be produced by a rise of several degrees in global temperatures. In fact, nuclear energy has a remarkable safety record in comparison with other forms of energy generation. The difficulties of disposing of nuclear waste have been greatly exaggerated. The commonly voiced statement that much nuclear waste remains dangerous for 100,000 years is pure nonsense. All matter is to some extent radioactive. Highly radioactive waste loses its radiation quite rapidly, levels falling to those found naturally on earth in 500, or at most 1,000 years. These are long periods, but safe storage for them should offer few problems. Then (as is shown in Chapter 12 of this book) the danger of low levels of radiation such as may arise from minor accidents at power stations have been greatly over-estimated. If the greenhouse effect impels us to introduce more safely operated nuclear power stations, as well as making us more economical in our use of natural resources, it will have done mankind a good turn.

So far we have been concerned with better ways of obtaining energy. Much of the carbon dioxide entering the atmosphere comes from cutting down forests (and burning the timber) and from ploughing and cultivating peat lands. Trees in most developing countries are the main source of fuel. This will not raise atmospheric levels of carbon dioxide if a sustainable yield of fuel wood is available as the result of planting sufficient trees to meet the demand. In this case, the carbon dioxide is simply recycled and higher levels are not produced. Extensive tree planting throughout the world will also have important effects. In developing countries where aid is provided from abroad, this could be provided only if a certain quota of trees is planted and properly maintained; if the trees are eventually burned there will be no permanent improvement to the atmosphere. Timber used for building retains all its carbon. It has been suggested that some wood should be buried, where, over millions of years, it may be transformed into coal. Unfortunately the alternative is that it will be emitted from the burial ground as methane, as happens in refuse landfills in Britain (see p. 35). If the methane can be collected and burned, it will produce carbon dioxide, which may appear to have foiled our objectives. However, as carbon dioxide has only half the greenhouse potential of methane, this is something which might be encouraged.

Is there any possibility of these measures being put into operation? I think that the answer is yes. The various strategies for economy are, in the long run, good business and worth implementing in any event. So are the increased uses of renewable energy. Fig. 25 shows that nearly three-quarters of the carbon dioxide at present emitted comes from Western Europe, North America, the USSR and its satellites, Japan and Australia. The substitution of nuclear power for that derived from fossil fuels in these countries and the use of as much as possible of that electricity for transport, industrial power and space heating would at least halve these nations' output of carbon dioxide. If developing countries, as they became more advanced, followed the same route, the problem of carbon dioxide would be more than solved.

Unfortunately there are the other greenhouse gases to consider. I think that the CFCs and oxides of nitrogen are likely to be controlled by measures already planned, and do not need further consideration. This leaves us with methane. The first essential is for more research, to determine just how much is actually being produced by ruminants and by fermentation in rice paddies, if these are indeed the main sources. If they are, control may be very difficult if not, in some instances, impossible, but some progress might be made. As meat is not essential for our diet, the number of cattle could be greatly reduced. Milking cows could receive hormonal injections to maximise milk yields, provided this also reduces methane output. The livestock problem should not be completely insoluble. We will need to look into ways in which rice is grown. Yields may be smaller, yet still adequate, if wet methanogenic paddies are abolished and if another variety of the grain is grown in drier conditions. The importance of marshes and swamps needs to be assessed, and we must perhaps face the possibility that some nature reserves in these situations may need to be sacrificed to prevent wider damage to the world. But in the long run it may be very difficult to control methane in the way carbon dioxide could be controlled. In this case we may need to be even more stringent in our measures against carbon dioxide to compensate for our failure to reduce sufficiently the output of methane.

Finally, it must be realised that not every scientist agrees with what has so far been said in this chapter. Academician Mikhael Budyko of the Soviet Union thinks that increased levels of carbon dioxide would be a good thing, and that instead of trying to reduce output this should be increased. He believes that, on balance, the world's climate will be improved. If one is mainly thinking of Siberia, this may well be true. But at a recent scientific congress in Hamburg, Germany, Academician Budyko received little support from workers in different countries.

Between 1850 and 1950 the world's mean temperature rose significantly, parallel to, but possibly not caused by, the rise in the levels of carbon dioxide and methane. For the next 20 or so years the rise was halted, and temperatures levelled off. At that time some meteorologists warned us that we were at the beginning of another Ice Age. They did not expect the conditions which obtained in Britain some 12,000 years ago to return in less than a few hundred or even thousands of years, but they thought we were in for a gradual cooling which would significantly affect all crops and wildlife. There are also those, especially the American John Hamaker, who have forecast a sudden drop in temperature with a further Ice Age before the beginning of the next century. It has been pointed out that during the last 2,500,000 years there have been 25 Ice Ages, each lasting about 90,000 years, and warmer periods of only about 10,000 years between the cold phases. There is some evidence to suggest that some of these Ice Ages arrived quickly, in not much more than 20 years. So we are at about the end of our warm period.

Perhaps the most comforting theory is that, were the levels of greenhouse gases not rising, we would in fact already be entering another Ice Age. As it is, this will be neutralised by the extra heat being retained by the greenhouse gases. This does suggest that future developments may enable us to control global temperatures. If we had vast stores of carbon dioxide which could be rapidly released, this could cause a rapid warming, and if we had some means of reducing carbon dioxide levels, this could have the opposite effect.

It is difficult to decide between all these theories. It is a field where there are no genuine experts who can claim to be true prophets. I prefer to rely on facts. Greenhouse gases have risen in their concentration in the atmosphere. With all the uncertainties regarding their possible effects, I believe we would be wise to try to prevent these levels from rising any further, particularly as all the measures proposed are in themselves valuable means of preserving our environment and our natural resources.

Chapter 11

THE OZONE LAYER AND ULTRA-VIOLET RADIATION

It is possible that terrestrial life only appeared on earth when an ozone layer developed in the stratosphere, some 10 to 50 km above the planet's surface. This ozone was important, as it trapped much of the ultraviolet radiation from the sun. This radiation, in sufficient doses, is harmful, and even lethal. There is concern at present that some human activities, and certain man-made chemicals, are destroying part of the ozone in the stratosphere, and that if this happens, it is feared that all living organisms on earth will suffer. Is it possible to estimate the magnitude of such a risk to our future?

The importance of ozone, the gas in which three atoms of oxygen combine (as O_3) to form a molecule, was considered in Chapter 6. Ozone is never present as more than a few parts per million of the atmosphere, but as it is a very unstable and reactive substance, its significance is much greater than its comparatively low concentration might suggest.

Ozone is, at ground level where we live, a dangerous pollutant which may be harmful to both plants and animals. It can be used in hospitals as a sterilising agent to kill harmful bacteria. In the 1930s there was a proposal to release quite high concentrations of ozone in the London underground railway system to kill 'germs' and safeguard the health of passengers. Fortunately this scheme never came to fruition, as effective levels of ozone would have done more damage to passengers than to the bacteria they were supposed to kill. There were other widely held views about the beneficial effects of ozone, which was believed to reinforce the bracing properties of the air in seaside resorts. In fact ozone levels are not higher there than elsewhere, and what was thought to be ozone was actually the smell of decaying seaweed.

It is worth repeating that in the atmosphere (i.e. from sea level to 10 km up) ozone concentrations are normally low, and that such levels do not appear to have any harmful effects. These background levels of ozone are the result of small amounts of gas which drift down from the stratosphere, and from ozone produced by the interaction of sunlight with various pollutants including oxides of nitrogen and unburnt hydrocarbons. When these substances are relatively plentiful (as, for instance, in cities with heavy motor traffic) and where there is strong sunlight, ozone levels soon reach a concentration which is phytotoxic. Britain is perhaps fortunate in seldom suffering ozone damage, because there is generally insufficient sunshine. At night the reaction is reversed, and ozone levels fall rapidly; this usually prevents the build-up of really high levels. Nevertheless we are probably right to consider ozone, in the troposphere where we live, to be a potentially harmful substance.

As already indicated, in the stratosphere ozone has a completely different action. It is only present in small amounts, three parts per million being quite usual, but this is essentially beneficial as it is sufficient to block the passage of a substantial part of the sun's ultraviolet radiation. This radiation might otherwise be harmful to all types of living organisms. In theory, the higher the level of ozone in the stratosphere the better, unlike conditions in the troposphere, where ideally the ozone level should be as low as possible.

We have a somewhat ambivalent reaction to ultraviolet radiation. It is easy to assume, as we lie on the beach of some Mediterranean resort, and expose our bodies to it, that it is doing us good, relying on the fact that the skin's protective action, combined with the production of the pigment melanin, which prevents some of the harmful radiation from penetrating too deeply, will result in an attractive and healthy tan. We know that ultraviolet light has some beneficial properties, in that it causes the production of vitamin D in our skin, but we have also been warned of the hazards of too much radiation, and the possible damage to the stratosphere's ozone layer from aerosol sprays, so we join with the Ambridge Women's Institute and the Prince of Wales in calling for a ban on these sprays, particularly those used to apply cosmetics or hair lacquer.

Too much ultraviolet can undoubtedly do real damage. Many sunbathers, after excessive exposure, suffer serious burns which may be hard to cure. Such burns may produce permanent scars. The commonest effect of excessive sun is skin cancer. Thus a high proportion of workers in sunny Queensland, Australia, who habitually work without their shirts, have developed this complaint. Normally skin cancer, which most commonly affects fair-skinned people, is non-malignant, it does not spread and it can be treated quite easily, but it is unpleasant and potentially disfiguring. More serious is the possibility that ultraviolet radiation also causes malignant melanoma, a very dangerous form of skin cancer, although the relationship is still rather uncertain.

Normal levels of ultraviolet are those under which most animals and plants have evolved, so it would be surprising if they caused any damage. It is possible that when a shade-loving plant, or a cave- dwelling animal, is exposed by man to unprotected radiation, damage will occur, but this is unlikely to happen in nature. Ultraviolet may well be one of the factors which helps to keep species in their proper places. This balance may be affected if the levels of ultraviolet radiation reaching the earth are altered.

The first scare relating to damage to the ozone layer arose in the 1960s when it was proposed to develop a great number of supersonic transport aeroplanes (SSTs) which, to achieve high speeds faster than sound, would have to fly in the less dense atmosphere of the stratosphere to minimise skin friction and to cut down fuel consumption. Until then only a few military aircraft had flown that high. There were fears that if hundreds of large planes were to fly there regularly, their engines would burn fossil fuel at a high temperature, oxides of nitrogen would be released, water vapour would also be produced, and this might seriously affect the stratospheric ozone. In the event only a few Concordes and the similar Soviet Tu-144 have gone into service, and they have not had any appreciable effect. Furthermore, new research and additional calculations suggest that, in modest quantities, nitrogen oxides are more likely to enhance than to deplete the ozone layer.

A further threat arose from agriculture. The increased use, worldwide, of nitrogen-based fertilisers could lead to an increased output of nitrogen oxides from the soil. This might migrate upward into the stratosphere and damage the ozone layer. Further studies of this problem suggested that the net effect of fertiliser use, and the production of oxides of nitrogen, would be a more rapid creation of ozone rather than depletion. It was then suggested nuclear bombs would also have a harmful effect on the stratosphere. Again research indicated that the net result would be a thickening rather than a thinning of the ozone layer – should anyone be left alive to appreciate it.

A much more serious effect has been feared from the use of a group of chemicals, the chlorofluorocarbons, or CFCs. The best known was invented by Du Pont de Nemours, and given the trade name of 'Freon'. These substances are made up of chlorine, fluorine and carbon in varying proportions. Some also contain hydrogen. With the present bad publicity the CFCs are receiving, an account of their properties, and their uses, may be surprising. The CFCs are extremely stable chemicals. It is impossible to set them alight, so they have proved to be useful fire extinguishers, and as such have saved many lives from chemical conflagrations which would otherwise have been difficult to control. They are completely non-toxic, for if ingested they do not decompose or react with any substances within the body. For many purposes, their properties make them ideal.

The CFCs include a series of chemicals with slightly different properties, which fit them for varied uses. Dichlorodifluoromethane, CCl_2F_2 (known also as CFC-12 or Freon-12) has long been used as a coolant in home refrigerators and automobile air conditioners. In the former instance it changes from the liquid to the gaseous phase at around room temperature, and as these changes occur they absorb or emit heat, and so cause the refrigerator to function. CFC-12 has the advantage that if the refrigerator springs a leak, there is no danger of fire or any harmful effect on health.

CCl_3F (CFC-11) has been used extensively as the propellant in aerosol spray cans. Its inert properties make it particularly useful when the spray is applied to the human body, either as a cosmetic or for more important medical purposes. These two substances (CFCs-11 and 12) have been produced in considerable quantities. Together with other CFCs used as cleaning agents – again valuable because of their stability so that they do not react with materials being cleaned – they have amounted to over 1 million tonnes a year from the USA alone.

Those who first used CFCs industrially never dreamed they might become global pollutants. When used as aerosols they clearly got into the atmosphere, but the quantities were small and, at low concentrations, were not expected to have any effect. Much larger amounts were sealed into refrigerators and air conditioners, and little of this was thought likely to escape. Unfortunately, in these days of built-in obsolescence most refrigerators have quite short lives, after which they go for scrap. Few companies try to salvage the Freon, which is usually allowed to evaporate and so finds its way into the atmosphere.

The CFCs have two properties which have made them a danger. First, their very stability means that once in the atmosphere they remain there for a long time, CFC-12 for an average of 80 years, CFC-11 for 40 years. This gives them plenty of time to find their way up from the troposphere into the stratosphere, where they remain. Secondly, although they are so stable under most circumstances, when exposed to ultraviolet light they release atomic chlorine. This takes part in a very complicated chain reaction which has the result after several steps of destroying ozone, two molecules of which become three of ordinary oxygen.

Although the possibility of CFCs affecting the ozone layer was recognised in the 1960s, it has proved difficult to determine how serious such destruction may be and, more importantly, how harmful the results are likely to be to humans and other organisms. A series of mathematical

THE OZONE LAYER AND ULTRA-VIOLET RADIATION 145

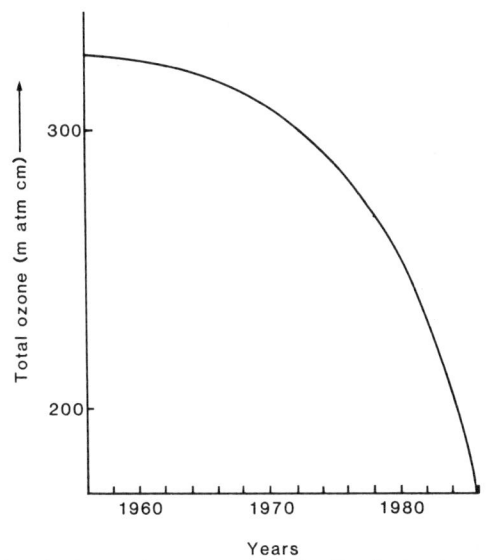

Fig. 27 The drop in the total ozone over Halley Bay in the Antarctic in the Octobers from 1957 to 1985.

models has given a predictably wide variety of results. Among other estimates, it was suggested in 1976 that if the present emission of CFCs continued, the ozone layer would be reduced by 6-7 per cent within 100 years. A later figure predicted even quicker results – 10 per cent reduction in 50 years. But as the thickness of the ozone layer fluctuates from day to day, often by amounts greater than those included in these calculations, the situation has remained very uncertain.

Recently, however, things have changed dramatically. J.C. Farman and the scientists of the British Antarctic Survey at Halley Bay on the Antarctic coast found that the levels of ozone, measured in October (Antarctic spring) from 1957 to 1984 fell from about 300 parts per thousand million to a figure of under 200 (Fig. 27). British, American and Japanese scientists have since made some even more startling discoveries. They have found that, following the dark Antarctic winter, massive ozone losses begin with the first sunlight of the spring in late August, and almost half of the total ozone disappears during September. Satellite observations show that the area of major ozone loss – colourfully described as the 'ozone hole' – is roughly the size of the United States of America. This is not a hole in the normal sense of the word. Fig. 27 indicated that the mean ozone level in the affected area was, in October 1960, just over 300 Dobson units; in 1983 it was about 200 Dobson units. So it would perhaps be more accurate to speak of it as a 'dent'. The loss of ozone, which starts in September, continues until mid-October. The strong air circulation in mid-November dilutes the depleted air with 'ordinary' air which has not had its ozone content reduced. The 'hole' persisted longer in November-December 1987, and there seems to be general agreement among those who are most involved with this problem that things are likely to get worse year by year, particularly if the levels of CFCs in the atmosphere continue to increase.

The discovery of the Antarctic hole in the ozone layer has had a profound effect on scientific and political opinion throughout the world. There is now fairly widespread, though by no means universal, agreement that CFCs do affect stratospheric ozone. So far only minor thinnings of the ozone layer have been found outside the Antarctic region, but these have been observed in several sites. The worry is that what is happening in Antarctica may be an earnest of what may happen globally. Consequently there have been several international conferences where most developed countries have agreed to restrict and eventually to ban the use of CFC compounds. For some of their uses, as in aerosols, there are already substitutes which are being increasingly used. New compounds for fire extinguishers and refrigerator coolants are not so easy to come by, but research will no doubt produce suitable substances before long. There is, however, always the danger that we may prematurely start to use other substances which have not been properly tested, and which may have even more harmful effects on man and the environment.

Not everyone is satisfied with the agreements by the major powers to restrict the production of CFCs. Some people want an immediate and total ban. They consider that the agreement to freeze production at present levels immediately, and to reduce it by 50 per cent in the 1990s, is not enough, since this will have little effect for some time, as the CFCs take several years to reach the stratosphere. It is possible that even if we stopped using CFCs today, there is enough on the way up to raise stratospheric levels considerably for some years to come.

It is still uncertain how the CFC ban will affect the general situation. There are other sources of atmospheric chlorine, for instance, from methyl chloride, which is released naturally or as the result of burning surface vegetation; and chloromethane, produced by wood-rotting fungi. These substances have been naturally produced for very long periods in quantities exceeding that of the CFCs, which makes one wonder why the ozone layer has survived at all. And whilst admitting that the Antarctic ozone hole is disturbing, we have no reason to suppose that the depletion will spread to affect the whole globe. Even if it does, we do not know how serious the result will be.

THE OZONE LAYER AND ULTRA-VIOLET RADIATION 147

This last point is the most important. Some people talk as if we were about to encounter a situation similar to that which prevailed in the early days of the world's history, when the atmosphere contained no oxygen, and therefore no ozone, so that any primitive life could only exist underwater where the lethal ultraviolet radiation could not penetrate. In my opinion there is no chance of this occurring. Probably the worst thing that could happen in temperate latitudes such as those of Britain, even if the ozone layer were to disappear completely, is that ultraviolet radiation levels would increase only to about those already existing in the tropics. If people were sensible, wore hats in summer and did not sunbathe too enthusiastically, they need not suffer. Only mad dogs and Englishmen are said to go out in the midday sun, yet inhabitants of the tropics have done so for many years without suffering any real damage. The higher levels which might occur in the tropics could have more serious effects, and might make it sensible for pale-skinned people to take more precautions, perhaps reintroducing the solar topee.

Some observations are encouraging. A violent solar storm in August 1972 produced a shower of protons which got rid of 16 per cent of the ozone layer. This must have let through an appreciable extra amount of radiation, but no biological effect was detected. The 1895 eruption of the volcano Krakatoa had various global effects, the best known being the dust cloud which reduced temperatures for many months. It also destroyed almost one-third of the ozone layer. Some of those concerned with CFC have suggested that such changes could cause much of the life on earth to perish. In the event, everything appeared to survive.

So I am not greatly concerned about the possible effect of CFCs on the ozone layer. Nevertheless there is *some* risk, and I would agree that we should take steps to avoid it. It may not be as easy as has been suggested to replace the CFCs with harmless substitutes, but it may be wise to try; in the long run, the expense should not be crippling. In my view, however, a much more important reason for phasing out the CFCs is their potential contribution to the greenhouse effect (see p. 129). So, if we act for the somewhat doubtful reason of defending the ozone layer, it will not be the first time we have done the right thing for the wrong reason.

Chapter 12

RADIOACTIVE WASTES

There is no doubt that high doses of ionising radiation kill, and that even moderately high doses cause cancer and other forms of sickness. It is equally certain that nuclear power stations produce considerable quantities of radioactive waste, and that under some circumstances the radiation from this waste could cause people or animals exposed to it to die or to develop cancers. For the purposes of this book, we need to know whether radiation, particularly what we may call 'waste radiation', has any ecological effects. While we should be particularly concerned with radioactive wastes, we are also interested in the discharge of radioactive substances from the nuclear industry, from accidents and from military uses of radioactive substances. Moreover, there are the effects of natural radiation, the so-called 'background' radiation to which we are all exposed, and with the effects of natural radiation which is present in different locations, but for which human activities bear no responsibility.

While I do not wish to get bogged down with considerations of theoretical physics, and discussions of the apparently complicated units used to describe different types and levels of radiation, the topic cannot be entirely ignored, if we are to understand the ways in which radiation may affect living organisms, and the possible effects of different intensities of radiation.

The subject is somewhat complicated. Radiation consists of electromagnetic waves or bits of atoms moving at high speeds, and these can damage living cells. Different radioactive substances produce different types of radiation, which have their own biological effects. Some radioactive substances give off rays, which affect the tissues outside the body, others produce energetic electrons, the so-called a rays, which only penetrate a few cell layers, and so are only effective when actually on, or in, the body. Finally, some produce particles, which are actually nuclei of helium atoms. These deposit a large amount of energy over a very short path in the tissues, and are generally the most damaging. Radioactive substances taken into the body may be deposited in sites (e.g. the thyroid gland) where they are in a position to do maximum harm.

Radioactive substances do not go on emitting radiation for ever, they all 'run down' eventually. Some of the radioisotopes produced in a nuclear explosion give out all their radiation in a fraction of a second. Some continue for thousands, even millions, of years. The concept of 'half life' is useful in this connection. Thus strontium- 90, which gives off a rays, has a life half life of 28 years. This means that at the end of 28 years the level of radiation given out falls to half, by 56 years to a quarter, and after 112 years to an eighth of the original level. Plutonium-239 has a half life of 24,000 years, and so is often said to be a very long-term hazard to the environment. However, the amount of radiation given out by a sample in a given time may be trivial. Uranium-238 has a half life of 4,470,000,000 years. This is roughly the age of the earth, so only half the uranium-238 that existed when the earth was formed has decayed away.

The many units used to describe radiation may seem confusing, but as we are concerned here with the biological effects of radiation on living organisms, the 'absorbed dose' is perhaps the most relevant conception. Until recently the unit most commonly used was the rad, which is still employed by many writers. In the S.I. system the Gray (Gy) is now used, and the conversion is simple: 1 Gy = 100 rads

The former unit of radioactivity was the Curie (Ci). This was a very large unit, and activity often had to be measured, for instance from the fall out of a distant bomb, in picocuries (pCi = Ci x 10^{-12}). The S.I. unit is the Becquerel (Bq) and 1 Bq = 2.7 x 10^{-11} Ci.

The Bq is thus a very small unit, and this may be confusing. After the Chernobyl accident, when lambs in Britain picked up some radiation, the British government forbade consumption of meat containing more than 1000 Bq of caesium-137 in a kilogram which seems quite a large amount. According to Professor J.H. Fremlin, however, anyone consuming a whole kilogram themselves would receive a dose which, in terms of carcinogenicity, would only be the equivalent of smoking a single cigarette.

The biological effect of radiation obviously depends on its intensity. Thus a really high dose of, say, 100 Gy will be immediately fatal. A substantial dose, in the order of 2-6 Gy, will cause burns and other acute effects, and some of those exposed will die within six weeks. We are mainly concerned with much lower doses, such as may have been experienced in this country, either as a result of leaks from nuclear power plants, fall out from testing of nuclear weapons, or as the result of such

150 WASTE AND POLLUTION

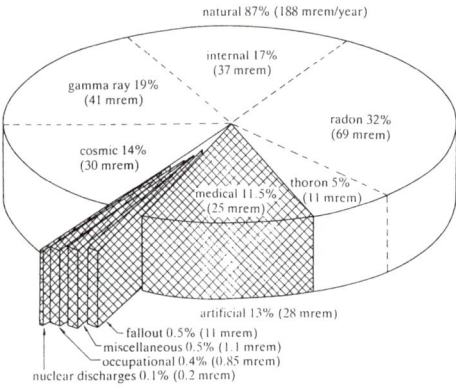

Fig. 28 Average proportions of background radiation from different sources in Britain (From Fremlin 1987).

incidents as the Chernobyl accident, when radioactivity from Russia was transported to Britain.

Before considering other sources, let us look at the background radiation to which we are all exposed. Fig. 28 shows that 87 per cent of the radiation comes from 'natural' sources, and only 13 per cent is man-made. The natural background radiation comes from rocks and from potassium within the body, as well as smaller amounts from various radioactive materials. The greater part of the man-made radiation is from medical sources, particularly X-rays, so this is very unevenly distributed. Some people, who never visit hospitals, will receive very little from this source. The amount arising from nuclear weapon fall out and from the discharges from power stations is tiny. The figure, incidentally, makes use of a unit, the millirem. This is a measure of the so-called *dose equivalent*, which allows for variations in the way different types of radiation (i.e. X-rays, gamma rays, electrons and alpha particles) affect living organisms. The background radiation fluctuates in different parts of the world, and even in different parts of Britain. The cause of the variation is mainly the type of rock. Granites are much the most radioactive, and thus the background radiation in Cornwall, or near Aberdeen, is much higher than in the rest of Britain. These contrasts may provide a useful clue as to the effect of low levels of radiation.

All types of radiation so far considered are called 'ionising radiations'. They have the property of knocking out electrons from the atoms in the substances through which they pass, and thus producing 'ionised atoms' which have great chemical activity. When this happens in a living cell, the usual effect is for the cell to be damaged. Thousands of cells in our bodies are damaged during the time it takes to read this chapter as we are continually bombarded with background radiation. Damaged cells may be killed, and then will be absorbed into the tissues; or they may be partially damaged, as with the DNA in the chromosomes. Such a damaged cell can be the focus of a cancerous growth. If the damage is to the germ cells, which produce sperm or eggs, these cells may, when they develop into recognisable organisms, differ to some extent from their parents; in other words, *mutation* has taken place.

Fortunately, the vast majority of the cells which are damaged by radiation are automatically repaired, and go on to lead a normal life. But a very small proportion are not repaired, and may cause cancer or mutations. The greater the extent of the radiation (assuming it is not sufficient to prove lethal) the greater the risk that a cell or cells will not be repaired. Older people develop cancer more commonly than younger members of the community, because over the years they have more cells damaged by radiation. As already stated, high levels of radiation may prove fatal, or at lower levels, very damaging to the tissues. What we need to know is the probable effect of exposure to *very* low levels.

At one time it was believed that low levels of radiation which produced no instant results were either harmless or actually beneficial. Consequently many workers were exposed to doses which would be considered very dangerous today, and some developed leukemia or other forms of cancer. The permitted dose has therefore been greatly decreased to levels which some people now consider unnecessarily low. We no longer permit the sale of such commodities as radioactive bath salts or mineral waters that claim to be the most radioactive in the world! It is probable that such products actually carried little or no radioactivity, but not so long ago such claims were widely believed. Nor is it impossible that in the future we may once more argue that low levels can be beneficial.

At present legislation and regulations are based on the linear hypothesis. We know approximately how dangerous a large dose is. The hypothesis assumes that damage will be proportional to the dose, and that there will be no threshold below which damage does not occur. This theory can be defended. The number of cells damaged must be proportional to the dose, and there is no reason to assume that where damaged cells are few they will be less likely to show signs of damage. Radiation levels that have been permitted have been where the number of cases of cancer are so small as to be unrecognisable, taking the normal rate of cancer in the population involved.

However, there is no proof that the linear hypothesis is correct. In Japan, the long-term radiological effect of the atomic bombs dropped at Hirohima and Nagasaki in 1945 was the production of fewer cancers

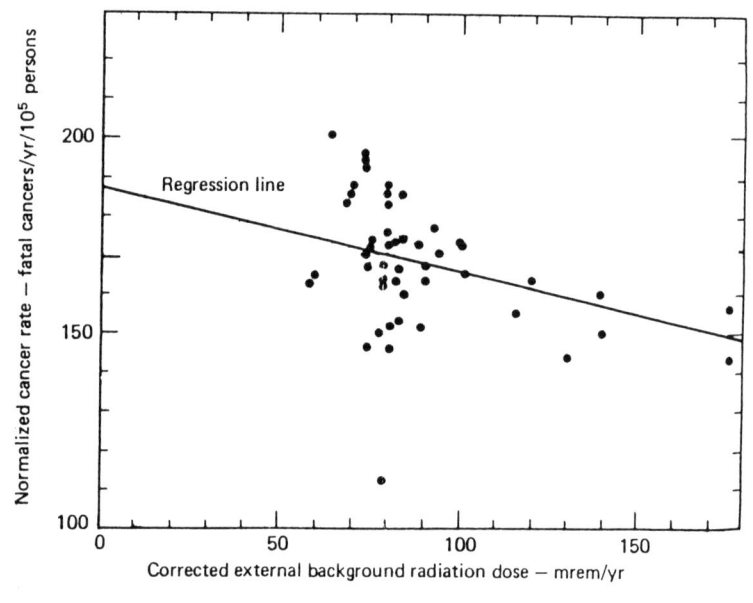

Fig. 29 Total cancer incidence against background radiation, USA (After Cohen 1980)

after a period of 45 than had been expected. Studies of the effects of the Chernobyl accident by the International Atomic Energy Agency in Vienna in 1991 found that, 5 years after the accident, there was no increase in any radiation-related illness, such as leukemia, birth defects or thyroid cancer in the villages near the reactor which had been contaminated by radiation.

In fact, there is growing evidence that low levels of radiation may produce fewer cancers than the linear hypothesis would predict, and even that radiation up to several times the background level may in some way be beneficial. It is well known that many poisons, lethal at high levels, may be beneficial in low doses, and may actually be used in medicine as drugs. There is now a considerable amount of evidence to suggest that the phenomenon of hormesis may operate in the case of ionising radiation. There is no general agreement on this but the observations of many who work in this field suggest that there is very little evidence that low levels of radiation are as harmful as is generally assumed.

The first general observation, based on investigation of various places with different background levels of radiation, is that nowhere has any area with high background radiation proved to have a higher than

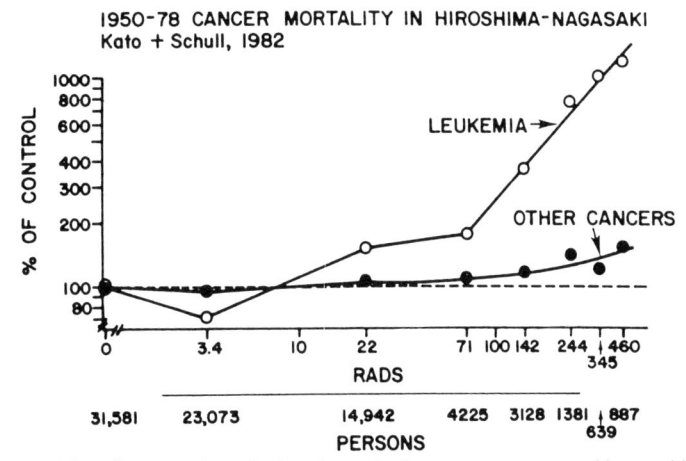

1950-78 CANCER MORTALITY IN HIROSHIMA-NAGASAKI
Kato + Schull, 1982

Fig. 30 Mortality rates from leukemia and other cancers among Nagasaki and Hiroshima survivors as a function of radiation received (From Luckey 1988)

average level of cancer. In the United States several surveys have shown the very opposite results, depicted in Fig. 29. It has been suggested that these results could be explained differently. The lowest level of cancer is in the state of Utah, where there is a high proportion of Mormons who do not drink or smoke. The other states with high natural radiation and below average cancer are mainly rural with comparatively little air pollution. So if this were an isolated observation, it would need to be treated with caution.

A second set of data concerns the effects of the radioactive gas radon, which bubbles up out of rocks such as the granite of Cornwall, and can build up to fairly high levels in houses with poor ventilation. B. Cohen of Pittsburgh, USA, has examined data from Finland, Sweden, China and no less than 415 counties in the USA. In all cases studied, the cancer levels were lowest in the areas with the greatest amount of radiation from the radon. These results challenge the pronouncement of the US government's surgeon-general, who said (with no hard evidence to support him) that in the USA radon was causing 20,000 deaths a year, and was the second most serious cause of the disease after cigarette smoking. Cohen does not say that radon never causes cancer, but he insists that these findings challenge the linear, no-threshold theory of cancer risk.

Another interesting result comes from Japan, from children who were under 10 years old at the time the atom bombs were dropped. Many of these received high doses of radiation. Those receiving above 100 rem showed, eventually, a considerably raised death level from cancer, of which about half were from leukemia (Fig. 30). The numbers are not high enough to be conclusive, but they do show that the children getting less than 100 rem had a cancer incidence below the background for Japan as a whole.

Some animal experiments support the same conclusion. Mice receiving low doses of a few rem in experiments made by E. Lorenz (low doses, but well above the background) generally lived longer than controls. Of course, more heavily exposed mice died in a shorter time. R.J. Pentreath found that salmon derived from irradiated eggs survived better than those whose eggs had not been so treated. D.J. Jefferies found that if the Saw-toothed Grain Beetle (*Oryzaephilus surinamensis*) is subjected to gamma radiation, pupae receiving up to 4 rads survived better than controls, though 10 to 20 rads proved fatal.

It is too early to be dogmatic about these results, but there is a growing volume of similar data, much of which was summarised in a special issue of the journal *Health Physics* in May 1978. There have also been experiments which have given negative results. It has been suggested by J.H. Fremlin that any radiation, even at very low doses, must initiate some cancers. However, this same radiation may also produce an improvement in the immune system, the way in which damaged cells are repaired. This would support the observations, and also explain why there may not always be a 100 per cent agreement between exposure to radiation and its beneficial, or its harmful, effects.

For many people, the greatest objection to nuclear power is that it produces dangerous radioactive waste. It is said by some, and widely believed, that we have no satisfactory means of disposing of this waste, which will remain dangerous to life for long periods. If this waste were to be treated as waste from the LeBlanc process was treated 100 years ago, i.e. simply dumped in any convenient waste site with no regard to anyone living in the vicinity, the results would be serious. But no one proposes to do this. There are very detailed plans for waste disposal, and the fact that they are not all operative today is due to uninformed opposition, often politically inspired.

The final waste from a nuclear power station, from the reactor core and everything which has come into contact with it, is radioactive, and emits α and β particles and gamma rays at levels which could be very dangerous. Wastes are classified as low, medium or high level, depending on the level of radioactivity. Low-level wastes are the least dangerous, and the easiest to dispose of. They have also been the cause of the greatest political uproar. Several sites in Britain were proposed for the burial of such low-level wastes. These substances would have been buried in trenches 9 m deep, and covered with 2 m of clay. The radiation reaching the surface from such a site filled with this waste has been calculated as less than a millionth of the natural potassium radiation reaching one person from another sitting in the same room. No α or β particles would pass through the clay cover. Even if someone fell into a filled trench, assuming all the most dangerous estimates or risk, the increased chances of developing cancer are so small as to be almost incalculable. On this subject, Professor Fremlin, one who has done much to combat the dangers from radiation, said: 'The callous cruelty of the irresponsible propagandists who have successfully terrified uninformed people in areas that might be used for storage has not been exceeded since the witch hunts of the 17th century.'

Medium- and high-level wastes would need to be stored in special containers and kept for a long period, until decay had rendered them harmless. The high level waste is very hot and intensely radioactive, and needs to be stored carefully, perhaps for as long as 50 years, until it can be classified as medium-level. Medium-level wastes must be treated with caution. Probably the best solution is to convert them to a rock-like substance called 'synrock', which could be stored in deep mines or even in the deep ocean where very little mixing with shallower waters takes place. Propagandists have claimed that these wastes will remain radioactive for perhaps as long as 100,000 years, and so will need to be contained for that period. There is no guarantee that global changes during such a period would not bring such material to the surface, even if deeply buried today. This argument is based on a misunderstanding. It is true that wastes would contain, among other substances, some plutonium-249 with a half life of 24,000 years. But the quantities are so small that even the most radioactive waste would, in 500, or at worst 1,000 years, be quite safe to handle, and would be emitting less radiation than naturally occurring rocks which are mined without endangering the miners. So if nuclear waste is a problem, it is one which the opponents of nuclear energy have exacerbated. Until now, in Britain, nuclear waste has not affected the environment sufficiently to have had any effect on our species or on the flora and fauna.

There have been a number of much publicised 'leaks' of radiation from nuclear power stations, and these have been reported in such a way as to cause the maximum public concern. In no case has any leak, except in a few cases within the station itself when operating staff have been affected, caused any radioactive pollution to reach levels higher than a fraction of the background found in the vicinity. Some low-level radiation has been discharged to the sea from several power stations. This has never been shown to damage any aquatic organism. Radioactive substances have been found to be accumulated in some living organisms,

156 WASTE AND POLLUTION

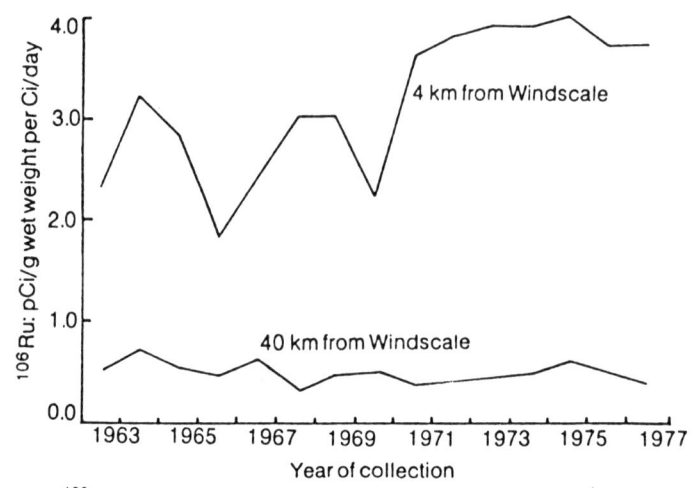

Fig. 31 ^{106}Ru in seaweed *Porphyra umbilicalis* from areas off Cumbrian coast, where low level discharge takes place (From Pentreath 1982)

[Graph: y-axis labeled "^{106}Ru: pCi/g wet weight per Ci/day" from 0.0 to 4.0; x-axis "Year of collection" from 1963 to 1977. Upper curve labeled "4 km from Windscale", lower curve labeled "40 km from Windscale".]

and concern has been voiced that if people ate these they might ingest a harmful quantity of radiation.

Two examples, which have had the effect of modifying the amount of material discharged from two power stations, illustrate what precautions have been taken. It was decided that no one who ate any of the organisms which picked up radioactivity should receive more than the equivalent of one-tenth of the background radiation from this source. From the Sellafield area the seaweed *Porphyra umbilicalis* is harvested, and made into 'laver bread' which is consumed in South Wales. Fig. 31 shows how ^{106}Ru was taken up by this seaweed; the levels of radiation near to the discharge were raised, but there was little effect further out to sea. Calculations were made as to the highest level of radiation if the most radioactive seaweed were harvested. The maximum amount eaten by anyone in Wales was also estimated, and these figures were used to calculate the level of radiation permitted in the power station's output.

In the case of discharge from the Winfrith generator, the lobster *Homarus gammarus* was the species most involved. Few people consume more than a few grams of this expensive crustacean in a year, but one man near Winfrith claimed to eat lobsters daily. This individual's unusual diet caused the permitted discharge to be substantially reduced. There was no evidence incidentally that the radiation had the slightest effect on the seaweed or on the lobsters, and in my opinion it would not have mattered much if their body levels of radiation had been much higher. Such radiation levels could hardly have had any harmful effects

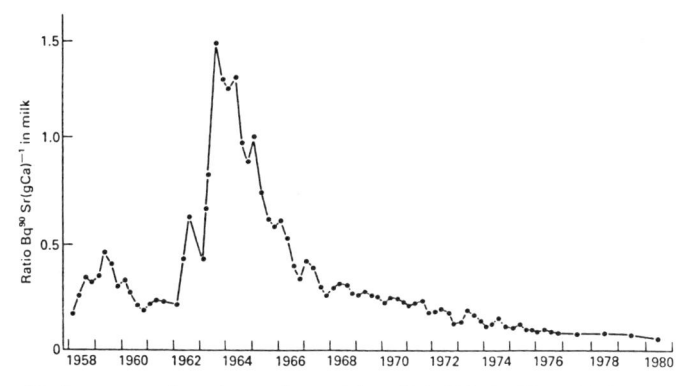

Fig. 32 Average ration of strontium-90 in milk in Britain (From Mellanby 1980).

Y-axis: Ratio Bq^{90}Sr(gCa)$^{-1}$ in milk

on the consumers even if laver bread and lobsters had been eaten in even greater amounts.

The Institute of Biology has been concerned with the possible dangers of radiation. As early as 1950 it organised a symposium on 'The Biological Hazards of Atomic Energy', which drew attention to possible dangers and how these might be avoided. Then in 1979 it held a second symposium, this time on 'The Biological Implications of an Expanded Nuclear Programme'. This was, to some extent, a stocktaking exercise. It showed that there was no evidence of any harmful, biological effect from the small amounts of radiation arising from nuclear power, and reached the general conclusion that an expanded programme need not pose any dangers provided existing standards of safety were maintained. Mention was also made of radioactivity from other sources, which confirmed that these were small compared with background levels. As Fig. 32 shows, in the 1960s, there was a significant rise in radioactive strontium in milk, as a result of testing bombs many thousands of miles away. The graph looks alarming, but the levels are still very low. When the Test Ban Treaty took effect, the levels fell and still remain virtually insignificant.

In recent years, the main concern has been caused by fall out from the Chernobyl accident, which was blown over to Britain, some of it being washed out in rain over the Welsh mountains and in the Lake District, where elevated levels were found in sheep and which remained in some animals even after two years. The British Government acted quickly, and banned the sale or consumption of sheep with more than 1000 Bq of caesium-137 in a kilogram of their meat. Although some compensation

158 WASTE AND POLLUTION

has been paid, this action has caused considerable hardship to many sheep farmers, as animals thought to be affected had almost no cash value. In my opinion, the Government's action was totally unnecessary. As mentioned above (p. 149), meat with 1000 Bq per kilo would, if anyone ate this amount, increase his or her risk of developing cancer as much as smoking one cigarette, but only if the linear hypothesis applied. Put another way, an individual flying the Atlantic would be exposed to a higher level of radiation by flying at 10,000 metres above sea level for seven hours.

The fall out from Chernobyl also affected Sweden and other parts of Scandinavia. When reindeer were found to be slightly radioactive, the Swedish Government had the animals slaughtered even though this upset the whole culture of the Swedish Lapps. The danger from this source was tiny (if indeed it was a risk at all), far less than permitting the Lapps to smoke tobacco. The Norwegians had the same problem. Very sensibly, they took no such action. There is no evidence that this has had any harmful effects.

One other cause for worry about radiation has not been discussed. This is that the germ cells may be affected, which may cause harmful mutations. Professor Fremlin has suggested that had the background radiation been lower, mutations would have been less frequent, and the human race as we know it would not have evolved by this time. The worry is that most mutations are harmful, and produce imperfect organisms which seldom survive in nature, although nowadays, were they human, we might succeed in keeping them alive. In theory, all radiation which reaches the developing germ cells could produce mutations, the number being, presumably, proportional to the intensity of the radiation. As the background radiation does not seem to be causing many mutations, any tiny increase will probably produce even fewer. There is little or no evidence of a great increase in mutations in Japanese survivors of the nuclear bombs at Hiroshima and Nagasaki, although some received considerable doses of radiation. In an experiment in the Brookhaven Forest in the USA, a source of radiation sufficient to kill all trees and vegetation within 50 metres has been running for many years. There is a fairly sharp cut-off where trees seem to grow normally, though they must be subject to levels of radiation many times those which would be tolerated near a nuclear power station or a waste dump. I know of no reports of great numbers of mutations in any of the animals which roam freely in this site. The concern about mutations from man-made nuclear sources appears now to be much less than it was 20 years ago.

My general conclusion must be that, in Britain today, nuclear power, the wastes from nuclear power stations, any accidental leaks from such stations, and any radiation from overseas deposited in Britain have so far had no recognisable effect on the human population, or even on wildlife. I believe that even with a vastly increased nuclear industry, if reasonable precautions are observed, there should be no significant effects in the future.

Chapter 13

THE FUTURE

Pollution that arises from wastes of all kinds is undoubtedly damaging our environment and harming our wildlife. Modern technology, directly or indirectly, is largely responsible for this damage. However, there are many examples of ways in which technology has been used to reduce damage, and to clean up the results of earlier insults to the environment. It is only by the proper use of technology that our future environment will be safeguarded.

It is nowadays fashionable to talk about an 'environmental crisis', and to lament the imminent destruction of the world's life support system. I wish to dissociate myself totally from this point of view. There are many things wrong with the way we manage our country and our planet, but there are also many things which are better today than at any other period for which we have any accurate information. At least in the more developed countries, a far larger population lives a healthier and more comfortable life than ever before – this is clearly demonstrated by the statistics showing the greater life expectancy as compared with only a few years ago. It is true that there are pockets of deprivation even in the most prosperous countries, and that starvation and famine can devastate some of the poorer nations, but we have the resources to overcome these situations if only we would use them. These are the resources provided by modern technology.

However, while I believe that our species, and the world's life support system, can be preserved, I think that the prospects for some of our wildlife, in particular the larger tropical mammals, are not very bright. This is because there are too many human beings living in the world and because of the way this human population is increasing. As long as we believe that every baby born has the right to a reasonably comfortable life, with adequate food and shelter, and as long as our numbers continue to increase, there must be less space left for wildlife. We must make sure that the space that remains is properly used.

Today we are suddenly bombarded with stories about the way the world and its climate may be being affected by pollution from man-made waste gases, and specifically the possible dangers of the greenhouse effect and the holes which are said to be appearing in the ozone layer. We are told that there are likely to be catastrophic changes in the weather, and that the risk of cancer from increased ultraviolet radiation is imminent. These are of course possibilities, but I am concerned at the way even scientists who should know better often fail to distinguish between theory and fact.

It is obviously better to be safe than sorry, so if there is a chance of dangerous developments, it is wise to try to take preventive action. As explained, I am not very concerned about the possible effects of damage to the ozone layer, but I do support the international moves to reduce and eventually eliminate the release of CFCs. This could be done without any serious ill-effects to our economy, and would be an interesting experiment in international cooperation on global issues.

The greenhouse effect is rather different. It cannot be stressed too strongly that we do not have the faintest idea whether there will be any effect on world climate, or when any changes will be noticeable. Changes, nevertheless, are possible, and these may have far more serious consequences than any damage from other pollutants. What we do know is that the levels of 'greenhouse gases' (particularly carbon dioxide and methane) are rising. I would support the idea that we should make a real effort to stem these rises, and at the same time we should do much more research on this subject. It should be possible to get the main industrial countries to agree to a programme to reduce carbon dioxide emissions, particularly as the programme is, for the most part, good sense anyhow. It includes greater economy in the use of energy, the use of as much renewable energy as possible (realising that solar power, wind power, etc. is never likely to produce more than a maximum of 20 per cent of our needs), and the substitution of nuclear power for fossil-fuelled electric generators. The electricity so produced could power an increased fraction of transport and industrial needs.

We have reduced the output of such pollutants as smoke and sulphur dioxide, so that urban dwellers generally enjoy improved conditions – and wildlife in towns flourishes. It is likely that before long we will reduce air pollution even further, with cleaner exhaust gases from cars (or by using electricity to drive them) and reduced output of pollutants from power stations, even from those using fossil fuels. So I believe that the effects of waste gases will be further reduced. Fresh water is not as clean as it should be, but this is because of a lack of effort, not a failure of technology. Here we need to distinguish real damage from the legitimate and harmless discharges which may, with proper safeguards, be made to

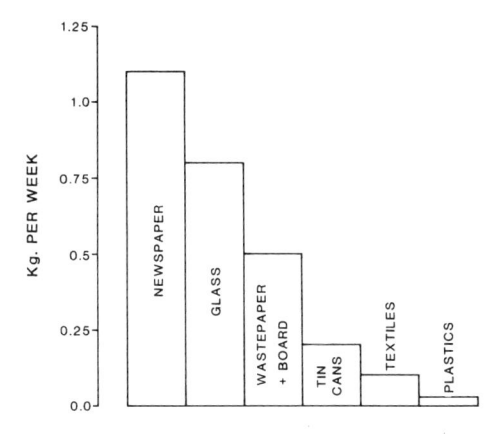

Fig. 33 Recyclable material produced per head in Britain each week.

rivers and the oceans. We need to be sure that our limited resources are used only where there is a real, and a soluble, problem.

Most modern factories are bigger, but also much cleaner, than those which operated a century ago. We must be sure that new industrial developments make the fullest use of technology to reduce their harmful effects on the environment.

As was shown in the earlier chapters of this book, much of the trouble with waste from homes and industry is that it takes up space which might otherwise have other and more important uses. So it would seem wise if we tried to minimise the amount of waste we produced. Many people have advocated a more vigorous programme of recycling wastes, of turning them into a 'resource'. Unfortunately, under British conditions, recycling has seldom paid. Collecting and sorting has cost more than the materials rescued are worth. Nevertheless I believe that this whole question needs to be re-examined. I have discussed waste paper, which nowadays is usually too expensive to collect. However, as it contributes substantially to the methane produced in a landfill, and prevents the site from being used for housing for many years, the loss to the site owner (which may well be the local authority) is far greater than the shortfall to the paper collector. For many other wastes, a realistic overall calculation would generally show that anything we could do to reduce the quantities of waste materials would be a good national investment.

The volume of waste, and the litter which disfigures our countryside and the streets of our towns, largely arises from the huge amount of unnecessary packaging used today. This could be greatly reduced without harming the consumer. Many foodstuffs do need to be protected, and here plastic film is invaluable, even if it is difficult to dispose of due to its 'non-biodegradability'. Plastics should only be used where they are needed, otherwise more easily disposable material should suffice. The mess in the countryside would be greatly decreased if cans were not used for drinks, and only returnable bottles (with a sufficient deposit as inducement) were permitted. Industry should be encouraged to make plans to reduce waste so as to protect the environment. In the long run it would even save money.

Several times in this book we have noted that wildlife may flourish under the most unpromising conditions. Waste tips have become SSSIs, sewage farms have attracted thousands of birds, and the most toxic wastes have provided conditions suitable for the most interesting flowers. Whereas I would not therefore suggest that we should take less care to contain our wastes, these developments may provide grounds for optimism. The world is tough and able to adapt to changed conditions. Whilst endeavouring, of course to avoid damage to valuable habitats, we should at the same time realise that a policy of positive conservation, which includes the creation of new habitats for ourselves as well as for wildlife, has much to commend it. Our wastes can then make a positive contribution to conservation, instead of always being a menace to the environment.

Appendix 1

FLOWERING PLANTS FOUND AT PITSEA LANDFILL SITE

Plants found at various surveys, 1975-82. None of these was intentionally introduced. A few non-native species, e.g. tomato and wheat, were found, but few unusual aliens appear to be present. Most long established waste disposal sites should have many of the species listed here.

Achillea millefolium, Yarrow
Aegopodium podagraria, Goutweed
Agropyron repens, Couch grass
Agrostis canina, Velvet bent
A. capillaris
A. stolonifera, Creeping bent
A. tenuis, Common bent
Alliaria petiolata, Hedge garlic
Alopocurus geniculatus, Marsh foxtail
A. myosuroides, Slender foxtail
A. pratensis, Meadow foxtail
Anagallis arvensis, Scarlet plimpernel
Anthemis cotula, Stinking mayweed
Anthriscus sylvestris, Cow parsley
Arctium lappa, Great burdock
Armoracia rusticana, Horseradish
Arrhenathrum elatius, Oat grass
Artemisia absinthium, Wormwood
A. vulgaris, Mugwort
Aster tripolium, Sea aster
Atriplex hastata, Hastate orache
A. littoralis, Grass-leaved orache
A. patula, Common orache
Avena fatua, Common wild oat

Ballota nigra, Horehound
Barbarea vulgaris, Yellow rocket
Bellis perennis, Common daisy
Beta vulgaris, Sea beet
Brassica juncea, Mustard

B. napus, Rape
B. nigra, Black mustard
B. oleraca, Wild cabbage
B. rapa, Annual turnip
Bromus hordeaceus, Soft brome
B. tectorum, Drooping brome

Calystegia sepium, Bindweed
Capsella bursa-pastoris, Shepherds purse
Carduus tenuiflorus, Slender-flowered thistle
Centaurea nigra, Knapweed
Centranthus ruber, Red valerian
Cerastium fontanum, Common mouse-ear
C. glomeratum, Sticky mouse-ear chickweed
Chamaenerion angustifolium, Rosebay willowherb
Chenopodium album, Fat-hen
C. bonus-henricus, Good King Henry
C. hybridum, Sow bane
C. polyspermum, Many-sided goosefoot
C. rubrum, Red goosefoot
Chrysanthemum leucanthemum, Ox-eye daisy
Cicuta virosa, Cowbane
Cirsium arvense, Creeping thistle
C. vulgare, Spear thistle

Conium maculatum, Hemlock
Coronopus squamatus, Swinecress
Crataegus oxycanthoides, Hawthorn
Crepis vesicaria, Beaked hawk's-beard

Dactylis glomerata, Cock's foot
Daucus carota, Wild carrot
Dipsacus fullonum, Common teasel

Elymus repens, Lyme grass
Epilobium hirsutum, Great willowherb
E. montanum, Broad-leaved willowherb
E. tetragonum, Square-stalked willowherb
Erophila verna, Whitlaw grass
Euphorbia helioscopia, Sun spurge
E. peplus, Petty spurge

Festuca rubra, Red fescue
Fumaria officinalis, Common fumatory

Galega officinalis, Goat's rue
Galium aparine, Cleavers
Geranium dissectum, Cut-leaved cranesbill
G. pratense, Meadow cranesbill
G. pusillum, Small-flowered cranesbill
G. robertianum, Herb robert
Glechoma hederacea, Ground ivy

Holcus lanatus, Yorkshire fog
Hordeum murinum, Wall barley
H. secalinum, Meadow barley

Kickxia spuria, Round-leaved fluellen

Lamium album, White dead nettle
L. amplexicaule, Henbane
L. purpurea, Red dead nettle
Lapsana communis, Nipplewort
Lathyrus nissolia, Grass vetchling
Lepidium campestre, Field pepperwort
Lolium perenne, Rye grass
Lotus corniculatus, Bird's-foot trefoil

L. tenuis, Narrow-leaved Bird's-foot trefoil
Lunaria tenuis, Honesty
Lycium chinense, Duke of Argyll's tea tree
Lycopersicon esculentum, Tomato

Malva sylvestris, Common mallow
Matricaria matricariodeds, Camomile
M. recicuta, Wild camomile
Medicago lupulina, Black medick
M. sativa, Lucerne
Melilotus altissima, Yellow melilot
M. officinalis, Ribbed melilot

Odontites verna, Red bartsia
Oenothera erythrosepala, Evening primrose

Phalaris canariensis, Canary grass
Phleum pratense ssp. *bertolonii*, Lesser cat's-tail
P. pratense ssp. *pratense*, Timothy
Picris echioides, Bristly ox-tongue
P. hieracoides, Hawk's-weed ox-tongue
Plantago lanceolata, Broadleaved plantain
P. major, Great plantain
Poa annua, Annual meadow grass
P. compressa, Flat-stalked meadow grass
P. pratensis, Meadow grass
P. trivialis, Rough meadow grass
Polygonum arenastrum, Knotgrass
P. aviculare, Common knotgrass
P. convolvulus, Black bindweed
P. persicaria, Common persicarea
Populus alba, White poplar
P. nigra, Black poplar
Potentilla reptans, Cinquefoil
Prunella vulgaris, Selfheal
Puccinella fasiculata f. *pseudo-distans*, Tufted salt-marsh grass

Ranunculus acris, Meadow buttercup
R. repens, Creeping buttercup
Reseda luteola, Mignonette

Rosa canina, Dog rose
Rubus fruticosa agg, Blackberry
Rumex crispus, Curled dock
R. maritimus, Golden dock
R. obtusifolius, Broad leaved dock

Sasgina procumbens, Procumbent pearlwort
Salix alba, White willow
Sambucus nigra, Elder
Scirpus maritimus, Sea club-rush
Senecio jacobaea, Ragwort
S. squalidus, Oxford ragwort
S. vulgaris, Groundsel
Silene alba, White campion
S. dioica, Red campion
Sinapis arvensis, Charlock
Sison amomum, Stone parsley
Sisymbrium orientale, Hedge mustard
Slanum dulcamara, Bittersweet
S. nigra, Black nightshade
Sonchus arvensis, Corn sowthistle
S. asper, Prickly sowthistle
S. oleracius, Common sowthistle

Stachys sylvatica, Wood woundwort
Stellaria media, Common chickweed
S. pallida, Lesser chickweed

Taraxacum officinale, Dandelion
Trifolium campestre, Hop trefoil
T. dubium, Lesser yellow trefoil
T. hybridum, Alsike clover
T. pratense, Red clover
T. repens, White clover
Tripleurosporum inodorum, Scentless camomile
Triticum aestivum, Wheat
Tussilago farfara, Coltsfoot

Urtica dioica, Stinging nettle

Verbascum thapsus, Common mullein
Veronica persica, Persian speedwell
Vicia cracca, Tufted vetch
V. sativa, Common vetch
V. tetrasperma, Small vetch

Appendix 2

ORNITHOLOGICAL RECORDS AT PITSEA MARSH, 1971–87

These records include rare vagrants seen on only one occasion (e.g. Scaup, Spotted Crake, Little Stint, Bar-tailed Godwit, Grey Phalarope, Roseate Tern, Nightjar, Green Woodpecker, Black Redstart, Red-backed and Great Grey Shrikes, Hawfinch, Ortolan Bunting), winter migrants present in varying numbers, and breeding residents. These records include sightings of birds in areas adjacent to the landfill site, but most if not all visited the landfill and many remained there.

Great Crested Grebe, *Podiceps cristatus*
Little Grebe, *Tachybaptus ruficollis*
Black-necked Grebe, *Pidoceps nigricollis*
Cormorant, *Phalacrocorax carbo*
Grey Heron, *Ardea cinerea*
Greater Flamingo, *Phoenicopterus ruber*
Mute Swan, *Cygnus olor*
Bewick's Swan, *Cygnus columbianus*
Whooper Swan, *Cygnus cygnus*
Emperor Goose, *Chen canagica*
White-fronted Goose, *Anser albifrons*
Greylag Goose, *Anser anser*
Canada Goose, *Branta canadensis*
Barnacle Goose, *Branta leucopsis*
Brent Goose, *Branta bernicla*
Shelduck, *Tadorna tadorna*
Widgeon, *Anas penelope*
Gadwall, *Anas strepera*
Teal, *Anas crecca*
Mallard, *Anas platyrhynchos*
Pintail, *Anas acuta*
Garganey, *Anas querquedula*
Shoveler, *Anas clypeata*
Pochard, *Athya ferina*
Tufted Duck, *Athya fuligula*

Scaup, *Athya marila*
Goldeneye, *Bucephala clangula*
Smew, *Mergus albellus*
Marsh Harrier, *Circus aeruginosus*
Hen Harrier, *Circus cyaneus*
Kestrel, *Falco tinnunculus*
Merlin, *Falco columbarius*
Red-legged Partridge, *Alectoris rufa*
Grey Partridge, *Perdix perdix*
Pheasant, *Phasianus colchicus*
Water Rail, *Rallus aquaticus*
Spotted Crake, *Porzana porzana*
Moorhen, *Gallinula chloropus*
Coot, *Fulica atra*
Oystercatcher, *Haematopus ostralegus*
Avocet, *Recurvirostra avosetta*
Little Ringed Plover, *Charadrius dubius*
Ringed Plover, *Charadrius hiaticula*
Golden Plover, *Pluvialis apricaria*
Grey Plover, *Pluvialis squatarola*
Lapwing, *Vanellus vanellus*
Knot, *Caladris canutus*
Sanderling, *Calidris alba*
Little Stint, *Calidris minuta*
Dunlin, *Calidris alpina*
Ruff, *Philomachus pugnax*

Jack Snipe, *Lymnocryptes minutus*
Snipe, *Gallinago gallinago*
Woodcock, *Scolopax rusticola*
Bar-tailed Godwit, *Limosa limosa*
Whimbrel, *Numenius phaeopus*
Curlew, *Numenius arquata*
Spotted Redshank, *Tringa erythropus*
Redshank, *Tringa totanus*
Greenshank, *Tringa nebularia*
Green Sandpiper, *Tringa ochropus*
Wood Sandpiper, *Tringa glareola*
Common Sandpiper, *Tringa hypoleucos*
Turnstone, *Arenaria interpres*
Grey Phalarope, *Phalaropus fulicarius*
Little Auk, *Alle alle*
Mediteranean Gull, *Larus melanocephalus*
Little Gull, *Larus minutus*
Lesser Black-backed Gull, *Larus fuscus*
Black-headed Gull, *Larus ridibundus*
Common Gull, *Larus canus*
Herring Gull, *Larus argentatus*
Glaucous Gull, *Larus hyperboreus*
Great Black-backed Gull, *Larus marinus*
Kittiwake, *Rissa tridactyla*
Sandwich Tern, *Sterna sanvicensis*
Roseate Tern, *Sterna dougalli*
Common Tern, *Sterna hirundo*
Black Tern, *Chlidonias niger*
Feral Pigeon, *Columba livia*
Stock Dove, *Columba oenas*
Woodpigeon, *Clumba palumbus*
Collared Dove, *Streptopelia decaocto*
Turtle Dove, *Streptopelia turtur*
Cuckoo, *Cuculus carorus*
Barn Owl, *Tyto alba*
Little Owl, *Athene noctua*
Long-eared Owl, *Asio otus*
Short-eared Owl, *Asio flammeus*
Nightjar, *Caprimulgus europaeus*
Swift, *Apus apus*
Kingfisher, *Alcedo atthis*
Green Woodpecker, *Picus viridis*
Great Spotted Woodpecker, *Dendrocopus major*
Skylark, *Alauda argensis*
Swallow, *Hirundo rustica*
Sand Martin, *Riparia riparis*
House Martin, *Delichon urbica*
Tree Pipit, *Anthus trivialis*
Meadow Pipit, *Anthus pratensis*
Rock Pipit, *Anthus spinoletta*
Yellow Wagtail, *Motacilla flava*
Grey Wagtail, *Motacilla cinerea*
Pied Wagtail, *Motacilla alba*
Wren, *Troglodytes troglodytes*
Dunnock, *Prunella modularis*
Robin, *Erithacus rubecula*
Nightingale, *Luscinia megarhynchos*
Black Redstart, *Phoenicurus ochruros*
Redstart, *Phoenicurus phoenicurus*
Whinchat, *Saxicola rubetra*
Stonechat, *Saxicola torquata*
Wheatear, *Oenanthe oenanthe*
Ring Ousel, *Turdus torquatus*
Blackbird, *Turdus merula*
Fieldfare, *Turdus pilaris*
Song Thrush, *Turdus philomelos*
Redwing, *Turdus iliacus*
Mistle Thrush, *Turdus viscivorus*
Cetti's Warbler, *Cettia cetti*
Grasshopper Warbler, *Locustella naevia*
Savi's Warbler, *Locustella luscubiodes*
Reed Warbler, *Acrocephalus scirpaceus*
Sedge Warbler, *Acrocephalus schoenobaenus*
Great Reed Warbler, *Acrocephalus arundinaceus*
Icterine Warbler, *Hippolais icterina*
Barred Warbler, *Sylvia nisoria*
Lesser Whitethroat, *Sylvia curruca*
Whitethroat, *Sylvia communis*
Garden Warbler, *Sylvia borin*
Blackcap, *Sylvia atricapilla*
Yellow-browned Warbler, *Phylloscopus inornatus*
Wood Warbler, *Phylloscopus sibilatrix*
Chiffchaff, *Phylloscopus collybita*
Goldcrest, *Regulus regulus*
Firecrest, *Regulus ignicapillus*
Spotted Flycatcher, *Muscicapa striata*
Pied Flycather, *Ficedula hypoleuca*

Bearded Tit, *Panurus biarmicus*
Long-tailed Tit, *Aegithalos caudatus*
Willow Tit, *Parus montanus*
Coal Tit, *Parus ater*
Blue Tit, *Parus caeruleus*
Great Tit, *Parus major*
Red-backed Shrike, *Lanius collurio*
Great Grey Shrike, *Lanius excubitor*
Jay, *Garrulus glandarius*
Magpie, *Pica pica*
Jackdaw, *Corvus mondula*
Rook, *Corvus frugilegus*
Carrion Crow, *Corvus corone corone*
Hooded Crow, *Corvus corone cornix*
Starling, *Sternus vulgaris*
House Sparrow, *Passer domesticus*
Tree Sparrow, *Passer montanus*

Chaffinch, *Fringilla coelebs*
Brambling, *Fringilla montifringilla*
Greenfinch, *Carduelis chloris*
Goldfinch, *Carduelis carduelis*
Siskin, *Carduelis spinus*
Linnet, *Acanthis cannabina*
Twite, *Acanthis flavirostris*
Redpoll, *Acanthis flamma*
Bullfinch, *Pyrrhula pyrrhula*
Hawfinch, *Coccothraustes coccothraustes*
Snow Bunting, *Plectrophenax nivalis*
Yellowhammer, *Emberiza citrinella*
Ortolan Bunting, *Emberiza hortilana*
Reed Bunting, *Emberiza schoeniclus*
Corn Bunting, *Miliaria calandra*

Appendix 3

DEFINITION OF SOME TYPES OF WASTES

Statutory definitions of some types of waste are given in the Control of Pollution Act 1974 (COPA). As will be seen, the situation is still confused, particularly when 'hazardous' or 'poisonous' materials are concerned. However, site licences for places where waste material may be deposited can be sufficiently precise to ensure that the environment is not seriously affected.

Household Waste (also Municipal Waste or Civic Amenity Waste)
Defined as 'waste from a private dwelling or residential home or from premises forming part of a university or school or other educational establishment or forming part of a hospital or nursing home'. Such waste may contain toxic and dangerous substances, though there are generally regulations as to what householders may put in their dustbins.

Industrial Waste
Waste from 'any factory within the meaning of the Factories Act 1961'. This may include toxic and hazardous substances, about which there are special regulations.

Notifiable Waste
Dangerous substances which had to be notified under the Deposit of Poisonous Waste Act 1972, and which then received appropriate treatment to prevent them doing damage. This act is now repealed, and the term is no longer used. These wastes are now 'Special Wastes' (see below) and 'Hazardous Wastes'.

Toxic and Dangerous Wastes
In Britain this phrase has no precise meaning; it is almost synonomous with 'Special Waste'. Confusion arises as a European Com-

munity Directive deals with what it calls 'Toxic and Dangerous Wastes'.

Special Waste
Waste which it is considered 'is or may be dangerous or difficult to dispose of that special provision is required for its disposal'. Details given in the appropriate regulations, based on COPA, makes it clear that special waste has the potential to cause acute harm or injury to persons directly exposed to it.

Hazardous Waste
Another imprecise term which includes Special Wastes and other wastes which have harmful effects on the environment, not just on human health.

Appendix 4

BIRDS AT RYE MEADS SEWAGE WORKS, HERTFORDSHIRE, 1957–62

Sewage treatment works, formerly called sewage farms, were wonderful sites for birds. Modern sewage works are more efficient, more hygienic and less of a nuisance to people living nearby, but they support little wildlife. This appendix is included to show what we have lost through this necessary improvement in public health engineering. (after T.W. Godwin)

Birds which bred on the site

Little Grebe
Mallard
Shoveler
Tufted Duck
Mute Swan
Red-legged Partridge
Partridge
Pheasant
Water Rail
Moorhen
Coot
Lapwing
Little Ringed Plover
Snipe
Redshank
Stock Dove
Woodpigeon
Collared Dove
Cuckoo
Little Owl
Tawny Owl
Kingfisher
Skylark
Swallow
Carrion Crow
Great Tit

Blue Tit
Wren
Song Thrush
Blackbird
Robin
Reed Warbler
Marsh Warbler
Sedge Warbler
Garden Warbler
Whitethroat
Willow Warbler
Chiffchaff
Dunnock
Meadow Pipit
Yellow Wagtail
Pied Wagtail
Starling
Greenfinch
Goldfinch
Linnet
Bullfinch
Chaffinch
Reed Bunting
House Sparrow
Tree Sparrow

176 APPENDICES

Visitors (recorded on more than 10 occasions)

Great Crested Grebe
Shag
Heron
Teal
Garganey
Widgeon
Scaup
Pochard
Goldeneye
Kestrel
Common Scoter
Ringed Plover
Golden Plover
Jack Snipe
Curlew
Green Sandpiper
Wood Sandpiper
Common Sandpiper
Dunlin
Willow Tit
Ruff
Great Black-backed
 Gull
Herring Gull
Common Gull
Black-headed Gull
Common Tern
Turtle Dove
Barn Owl

Swift
Green Woodpecker
Great Spotted Woodpecker
House Martin
Sand Martin
Rook
Jackdaw
Magpie
Jay
Long-tailed Tit
Nuthatch
Treecreeper
Mistle Thrush
Fieldfare
Redwing
Wheatear
Stonechat
Whinchat
Blackcap
Lesser Whitethroat
Spotted Flycatcher
Tree Pipit
Grey Wagtail
Siskin
Redpoll
Brambling
Yellowhammer
Corn Bunting
Black Tern

Vagrants (recorded on fewer than 10 occasions)

Black-throated diver
Slavonian Grebe
Black-necked Grebe
Cormorant
Bittern
Gadwall
Pintail
Red-crested Pochard
Goosander
Smew
Shelduck
White-fronted Goose
Canada Goose
Bewick's Swan
Buzzard

Sparrowhawk
Hobby
Peregrine
Merlin
Broad-billed Sandpiper
Grey Plover
Turnstone
Great Snipe
Whimbrel
Black-tailed Godwit
Spotted Redshank
Knot
Little Stint
Temminck's Stint
Pectoral Sandpiper

Curlew Sandpiper
Grey Phalarope
Red-necked Phalarope
Stone Curlew
Little Gull
Short-eared Owl
Lesser Spotted Woodpecker
Sanderling
Snow Bunting
Wryneck
Woodlark
Hooded Crow
Coal Tit
Marsh Tit

Bearded Tit
Redstart
Black Redstart
Nightingale
Grasshopper Warbler
Melodious Warbler
Wood Warbler
Goldcrest
Firecrest
Pied Flycatcher
Rock/Water Pipit
Great Grey Shrike
Twite

Bibliography

ADAS Farm Waste Unit (1985) A report on the 1984 survey of water pollution incidents caused by farm wastes in England and Wales. Ministry of Agriculture, Fisheries and Food.

Addiscott, T. (1988) Farmers, fertilisers and the nitrate flood. *New Scientist*, 120, 50-54

Addyman, P.V. (1974) Excavations in York 1972-73. First interim report. *Antiquaries Journal 54*, 200-231

Alexander, W.B. (1955) Birds on Sewage Farms. *The Advancement of Science, 12*, 113-114

Allaby, M. (1986) *Ecology Facts*, London, Hamlyn.

Alleman, J.E. and Kavenagh, J.T. Eds. (1982). Industrial Waste. Proceedings of the 14th Mid-Atlantic Conference. Ann Arbor Science.

Anon. (1986) Acidity in United Kingdom Fresh Waters. United Kingdom Acid Waters Review Group Interim Report. London, Department of the Environment.

Anon (1986) Climate variations and environmental impact. *Nature and Resources*, Unesco, 22, 3-5.

Anon (1986) What future for nuclear power? *Nature, 321*, 367-368.

Ashmore, M.N., Bell, J.N.B. and Garretty, C. (1988) Acid Rain and Britain's Natural Ecosystems. London, Imperial College Centre for Environmental Technology.

Benton, C., Khan, F., Monahan, P., Richards, W.N. and Sheddon, C.B. (1983) The contamination of a major water supply by gulls (*Larus sp.*) Water Res. 17, 789-798.

Bhatt, H.G., Sykes, R.M. and Sweeney, T.L., Eds. (1985) *Management of Toxic and Hazardous Wastes*, Chelsea, Michigan, Lewis Publishers.

Binns, W.O. (1984) Acid rain and forestry. Forestry Commission Research and Development Paper 134. Forestry Commission Edinburgh.

Binns, W.O,, Redfern, D.B., Renolds, K. and Betts, A.J.A. (1985) Forest health and air pollution: 1984 survey. F.C.R. and D. Paper 142, Forestry Commission, Edinburgh.

Binns, W.O., Redfern, D.B., Boswell, R. and Betts, A.J.A. (1986) Forest health and air pollution: 1985 survey. F.C.R. and D. Paper 147, Forestry Commission, Edinburgh.

Blowers, A. and Pepper D. (1988) *Nuclear Power in Crisis*, London, Croom Helm.

Bolin, B., Doos, B.R., Jager, J. and Warruck, R.A. (1986) The Greenhouse Effect. Climatic Change, and Ecosystems. *Scope 29*, Chichester, Wiley.

Brackley, P. (1987) Acid Deposition and Vehicle Emissions: Eurpean Environmental Pressures on Britain. Energy Papers No. 22, Joint Energy Programme, Royal Institute of International Affairs.

Bradshaw, A.D. and Chadwick, M.J. (1980) *The Restoration of Land* Oxford, Blackwell.

Bromley, J. and Parker, A. (1979) Methane from landfill sites. International Environment and Safety. 9-11.

Butterfield, J., Coulson, J.C., Kearsey, S.V. and Monahan, P. (1983) The herring gull *Larus argentatus* as a carrier of salmonella. J.Hyg. 91, 429-436.

Cape, J.N., Paterson, I.S., Wellburn, A.R., Wolfenden, J., Melhorn, H., Freer-Smith, P.H. and Fink, S. (1988) *Early Diagnosis of Forest Decline*. Institute of Terrestrial Ecology, Penicuik.

CEGB Research, No. 20, August 1987. *Acid Rain*, a special issue, contains the following papers: Chester, P.F. Foreword Crane, A.J. and Cocks, A.T. The transport, transformation and deposition of airborne emissions from power stations. Skeffington, R.A. Soil and its responses to acid deposition. Brown, D.J.A. Freshwater acidification and fisheries decline. Roberts, T.M. Effects of air pollutants on agriculture and forestry. Manning, M.I. Effects on structural materials. Chester, P.F. Acid rain - a prognosis.

Cope, C.B., Fullger, W.H. and Willetts, S.L. (1983) *The Scientific Management of Hazardous Wastes*. Cambridge, CUP.

Cottrell, A. (1981) *How Safe is Nuclear Energy?* London, Heinemann

Coulson, J.C., Butterfield, J. and Thomas, C. (1983) The herring gull *Larus argentatus* as a likely transmitting agent of *Salmonella montevideo* to sheep and cattle. *J. Hyg. 91*, 437-443.

Coulson, J.C. and Butterfield, J. (1985) Studies of a colony of colour-ringed Herring Gulls, *Larus argentatus*: Colony occupation and feeding outside the breeding season. *Bird Study 33*, 55-59.

Coulson, J.C., Butterfield, J., Duncan, N. and Thomas, C. (1987) Use of refuse tips by adult British Herring Gulls during the week. *J. Appl. Ecol., 24*, 789-800.

Darlington, A. (1969) *Ecology of Refuse Tips*. London, Heinemann.

Davies, A.W. (1984) Disposal of radioactive wastes. *Chemistry and Industry*, 3 Sept.

Davis, B.N.K. (1977) Chalk and Limestone Quarries as Wildlife Habitats. *Miner. Environ. 1*, 48-56.

Davis, B.N.K. ed (1981) Ecology of Quarries. The importance of natural vegetation. Institute of Terrestrial Ecology, Cambridge.

Denness, B. (1987) Sea-Level Modelling: The Past and the Future *Prog. Oceonog. 18*, 41-59.

Department of the Environment (1971) Refuse Disposal: Report of the Working Party of Waste Disposal. London, HMSO.

Department of the Environment (1983). Agriculture and Pollution: the Government response to the Seventh Report of the Royal Commission on Environmental Pollution. Pollution Paper 21, HMSO.

Derwent, R.G. and Kay, P.J.A. (1988) Factors Influencing the Ground Level Distribution of Ozone in Europe. *Envrion. Pollut.* 55, 191-219

Dix, H.M. (1981) *Environmental Pollution*, Chichester, Wiley.

Doxat, J. (1977) *The Living Thames: The Restoration of a Great Tidal River*. London, Hutchinson, Benham.

Ellsaesser, H.W. (1988) The Greenhouse Effect: Science Fiction? *Consumers Research, 71*, Special Report, 27-31

Ellsaesser, H.W., MacCracken, M.C., Walton, J.J. and Grotch, S.L. (1987) Global Climatic Trends as Revealed by the Recorded Data. *Reviews of Geophysics, 24*, 745-792.

Everest, D. (1988) *The Greenhouse Effect. Joint Energy Programme* Royal Institute of International Affairs.

Farman, J. (1987) What hope for the ozone layer now? *New Scientist*, 12 November, 51-54.

Fennell, H., James, D.B. and Morris, J. (1974) Pollution of a storage reservoir by roosting gulls. *Water Treatment and Examination, 23*, 5-20.

Ferry, B.W., Baddeley, M.S. and Hawksworth, D.L. eds. (1973) *Air Pollution and Lichens*, London, Athlone Press.

Fitter, R.S.R. (1945) *London's Natural History*, London, Collins New Naturalist.

Fremlin, J.H. (1987) Nuclear radiation - friend and foe. London, The Royal Society.

Gemmell, R.P. (1977) Colonization of Industrial Wasteland, *Studies in Biology No. 80*, London, Edward Arnold.

Gemmell, R.P. (1979) Design and management of industrial derelict sites. 189-204. From *Ecology and Design in Amenity Land Management*, Wye College.

Gemmell, R.P. (1981) The reclamation of acidic colliery spoil. *J. Appl. Ecol. 18*, 879-887

Gemmell, R.P. (1982) The origin and botanical importance of industrial habitats. From *Urban Ecology*, Oxford, Blackwell.

Gemmell, R.P. (1988) Tip top treasures. *Natural World, No.23* 7-9.

Gilbert, O. (1970) Further studies on the effect of sulphur dioxide on lichens and bryophytes. *New Phytologist, 69*, 605-627.

Gilbert, O. (1974) An air pollution survey by school children. *Environmental Pollution, 6*, 178-180.

Gilbert, O. (1987) The wildlife of Britain's wasteland. *New Scientist*, 24 March, 824-829.

Gibbs, A. (1961) The Bird Population of Rubbish Dumps. *London Bird Report, No. 26*, 104-110.

Goode, D. (1986) *Wild in London* (A Shell Book). London, Michael Joseph.

Gooders, J. and Lambert, T. (1982) *Collins British Birds*, London Collins.

Green, G.H. (1978) Worcestershire and Warwickshire Wintering Gull Project Second Progress Report for the period April 1977 to April 1978. Progress Report No. 2.

Greenwood, E.F. and Gemmell, R.P. (1978) Derelict industrial land as a habitat for rare plants in S.Lancs and W.Lancs. *Watsonia, 12*, 33-40.

Gregory, S. (1988) *Recent Climatic Changes*. London, Behaven Press.

Greig, J. (1981) The Investigation of a Medieval Barrel-Latrine from Worcester. *J. Archeol. Sci. 8*, 265-282

Greig, S.A., Coulson, J.C. and Monaghan, P. (1983) Age-related differences in foraging success in the herring gull (*Larus argentatus*). *Anim.Behav. 31*, 1237-1243.

Gribbin, J. (1988) The Ozone Layer. *New Scientist*, 5 May, 1-4.

Hawksworth, D.L. and Rose, F. (1975) Lichens as Pollution Monitors. *Studies in Biology*, London, Arnold.

Hazardous waste disposal and the reuse of contaminated land. Papers presented at a symposium organised by the Road and Buildings Materials Group, Industrial Health and Safety Group and the Environmental Group of the Society of Chemical Industry, 1984, all papers published in *Chemistry and Industry*, Sept.1984. Hillman, T.C. The waste disposal authority role. Cheney, A.C. Experience with the co-disposal of hazardous waste with domestic waste. Cook, J.D. Landfill as a disposal route for difficult wastes. Aldred, J.B. and Lord, D.W. Experiences in the investigation of contaminated land. Beckett, M.J. and Simms, D.I, The development of contaminated land.

Helder, T., Stutterheim, E. and Olie, K. (1982). The toxicity and toxic potential of fly ash from a municipal incinerator assessed by means of a fish early life stage test. *Chemosphere, 11*, 965- 972

Henderson-Sellers, B. (1984) *Pollution of our Atmosphere*. Bristol, Hilger.

Holmes, J.R. (1981) *Refuse Recycling and Recovery*. Chichester, Wiley.

Horton, N., Brough, T. and Rochard, J.B.A. (1983) The importance of refuse tips to gulls wintering in an inland area of south-east England. *J.Appl. Ecol.*, 20, 751-765.

Hynes, H.B.N. (1966) *The Biology of Polluted Waters*. Liverpool University Press.

Ineson, P. (1986) Pollution in Cumbria. Institute of Terrestrial Ecology, Merlewood Research Station, Cumbria.

Innes, J.L. (1987) Air Pollution and Forestry. *Forestry Commission Bulletin 70*, London HMSO

Innes, J.L. and Boswell, R.C. (1987) Forest Health Surveys 1987. Part 1, Results. *Forestry Commission Bulleton 74*, London HMSO.

Innes, J.L. and Boswell, R.C. (1987) Forest Health Surveys 1987. Part 2, Analysis and Interpretation. *Forestry Commission Bulletin 79*, London HMSO.

Jones, K.B.C. (1968) Farm Waste Management - Lessons From America. *Agriculture 75*, 213-218.

Joyce, Christopher (1988) Radon and lung cancer: the link is missing. *New Scientist, 29*, 29 September

King, T. (1981) Spray bath treatment cleans up liquids. *Surveyor*, 12 March

Krug, E.C. and Frink, C.R. (1983) Effects of Acid Rain on Soil and Water. *Bulletin 811*, Connecticut Agricultural Experiment Station.

Langford, T.E. (1983) Electricity Generation and the Ecology of Natural Waters. Liverpool University Press.

Langford, T.E. (1989) The Ecological Effects of Thermal Discharges on Aquatic Habitats. London, Elsevier ASP.

Leas, G. (1986) Acid Rain. *Observer* magazine, 19 October.

Lightowlers, P. (1988) A poisoned landscape gathers no moss. *New Scientist*, 5 May, 54-58.

Lonsdale, D. (1986) Beech Health Study 1985. F.C.R. and D. *Paper 146*, Forestry Commission, Edinburgh.

Lonsdale, D. (1987) Beech Health Study 1986. F.C.R. and D. *Paper 149*. Forestry Commission, Edinburgh.

Luscombe, G. (1988) Wild not derelict. *Natural World No.23*, 6-7.

Mann, E. (1935) Notes on the Birds of Edmonton Sewage Farm, 1933- 34. London Naturalist.

Mellanby, K. (1967) *Pesticides and Pollution*. London, Collins New Naturalist.

Mellanby, K. (1971) *The Mole*. London, Collins New Naturalist.

Mellanby, K. (1974) A water pollution survey, mainly by British school children. *Environmental Pollution, 6*, 159-173.

Mellanby, K. (1980) The Biology of Pollution, 2nd Edition. *Studies in Biology*, London, Arnold.

Mellanby, K. (1981) *Farming and Wildlife*, London, Collins New Naturalist.

Mellanby, K. (Ed.) (1988) Air pollution, Acid Rain and the Environment. Watt Committee on Energy, *Report No. 18*, Elsevier ASP.

Menser, H.A., Winant, W.M. and Bennett, O.L. (1979) Spray irrigation - a land disposal practice for decontaminating leachate from sanitary landfills. *ARR, NE 4*, 249–260. US Department of Agriculture.

Menser, H.A., Winant, W.M., Bennett, O.L. and Lundberg, P.E. (1979) The utilisation of forage grasses for decontamination of spray-irrigated leachate from a municipal sanitary landfill. *Environ. Pollut. 19.*

Monaghan, P. (1980) Dominance and dispersal between feeding sites in the Herring Gull (*Larus argentatus*). *Anim. Behav. 28*, 521-527.

Monaghan, P., Shedden, C.B., Ensor, K., Fricker, C.R. and Girdwood, R.W.A. (1985) Salmonella carriage by herring gulls in the Clyde area in relation to their feeding ecology. *J. Appl. Ecol., 22*, 669-680.

Moss, B. (1983) The Norfolk Broadland experiments in the restoration of a complex wetland. *Biol. Rev. 58*, 521-561.

Moss, B. (1987) The Broads, *Biologist, 34*, 7-13.

Moss, B. (1988) The palaelolimnology of Hoveton Great Broad, Norfolk: clues to the spoiling and restoration of Broadland. From *The Exploitation of Wetlands*, B.A.R., Oxford.

Moss, B., Balls, H., Booker, I., Manson, K. and Timms, M. (1984) The River Bure, United Kingdom: Patterns of change in chemistry and phytoplankton in a slow-flowing river. *Vehr. Internat. Verein. Limnol., 22*, 1959-1964.

Moss, B., Balls, H., Booker, I., Manson, K. and Timms, M. (1988) Problems in the Construction of a Nutrient Budget for the R. Bure and its Broads (Norfolk) prior to its Restoration from Eutrophication. From: *Algae and the Aquatic Environment*, Bristol, Biopress.

Moss, B., Irvine, K. and Stansfield, J. (1988) Approaches to the restoration of shallow eutrophic lakes in England. *Verh. Internat. Verein. Limnol., 23*, 414-418.

Mudge, G.P. and Ferns, P.N. (1982) The feeding ecology of five species of gulls (Aves: Larini) in the inner Bristol Channel. *J.Zool.Lond. 198*, 497-510.

Murton, R.K. (1971) *Man and Birds*, London, Collins New Naturalist

Parry, G.D.R. and Brummage, M.K. (1981) Solid Wastes: Reclamation and Management. *Landscape Research, 6*, 15-18.

Pentreath, R.J. (1981) The biological availability to marine organisms of transuranium and other long-lived nucleids. From *Impact of Radionuclide Releases into the Marine Environment*, International Atomic Energy Agency, Vienna.

Pentreath, R.J. (1982) Principles, practice and problems in the monitoring of radioactive wastes disposed of into the marine environment. *Nucl. Energy 21*, 235-244.

Porteus, A. (1981) *Refuse Derived Fuels*, London, Applied Science Publishers.

Reed, L. (1972) *An Ocean of Waste*. London, Conservative Political Centre.

Roberts, R.D. and Roberts, T.M. Eds. (1984) *Planning and Ecology* London, Chapman and Hall.

Rodgers, C. (1988) Global ozone trends reassessed. *Nature, 332*, 201.

Royal College of Physicians. (1970) *Air Pollution and Health*. London, Pitman.

Royal Commission on Environmental Pollution. Reports No. 1 (1771), No. 2 (1972), No. 3 (1972), No. 5 (1976), No. 7 (1979), No. 10 (1984) are all relevant.

Royal Commission on Environmental Pollution (1985). Eleventh Report: Managing Wastes - The Duty of Care. Cmd. 9675 HMSO.

Sagan, L.A. Ed. (1987) Radiation Hormesis. Special Issue of *Health Physics*, 52, 517-680.

Shuckley, M. (1981) *Environmental Archeology*, London, Allen and Unwin.

Simms, E. (1975) *Birds of Town and Suburb*. London, Collins.

Sitwell, N. (1983) Acid rain and tidal waves. *Spectator*, 26 Feb., 14-15.

Sterritt, R.M. and Lester, J.N. (1980) The value of sewage sludge in agriculture and effects of the agricultural use of sludges contaminated with toxic elements: a review. *Science of the Total Environment, 16*, 55-90.

Suess, M.J. Ed. (1985) Solid waste management: selected topics. Copenhagen WHO.

Sutton, C. (1988) Inside science - radioactivity. *New Scientist*, 11 Feb., 1-4.

Toke, D. (1988) Forecast: Drought and Drowning. *Environment Now, 9*, 30-33.

Valroff, J. (1985) *Pollution atmosphérique et pluies acides*. Paris, La Documentation Francais.

Vogler, J. (1978) *Muck and Brass*. London, Oxfam.

Whelan, C.D., Monaghan, P., Girdwood, R.W.A. and Fricker, C.R. (1988) The significance of wild birds (*Larus* sp.) in the epidemiology of campylobacter infections in humans. *Epidem.Inf. 101*, 259-267.

Wilson, D.C., Smith, E.T. and Pearce, K.W. (1981) Uncontrolled hazardous waste sites: a perspective of the problem in the UK. *Chemistry and Industry*, 3 January.

Index

As a rule only English or Common names are given for animals and plants. Latin names are only given for species with no English names. Species only included in the appendices are not given in the index.

Aberfan 44
Acid rain 14, 78, 88, 103
Aedes aegypti 125
Aedes punctor 125
aerobic decomposition 32
air pollution 69
Alder 37
alpha particles 150
Angelica 65
Annual Meadow Grass 39
Anglo-Saxon York 11
Ash 12
Aspen 13
Atrazine 14
Avocet 103

background radiation 145
Badger, 41, 118
Barnacle 127
Batterbee, R.W. 106
Bavaria 80
Beaked Hawksbeard 48
Bearded Tit 20
Bedfordshire and Huntingdonshire Wildlife Trust 50
Beech tree 81
benomyl 115
Bevan, R.J. and Greenhalgh, G.W. 74
Biochemical oxygen demand (BOD) 88
Birch 37
Bird's-foot Trefoil 53
Black Forest 80

Black-headed Gull 25
blackspot fungus disease 74
Bloodworm 91
"blue baby" syndrome 101
Blue-eyed Grass 65
Blueprint for Survival 8
body louse 117
Bordeaux mixture 114
Boron 52
Boyd, A.W. 103
Bradshaw, A.D. 43, 53
Bramble 67
Bridgewater Bay NNR 21
Bristly Ox-tongue 39
Bristol Channel 27
British Antarctic Survey 145
British Lichen Society 76
Brown Trout 108
Buckinghamshire 50
bunt 114
Bush Grass 8
Buzzard 118

Caddis fly 91
Camarthen County Council 22
carbon dioxide 129
carbon monoxide 85
carboxyhaemoglobin 85
Carrion Crow 25
cats, feral 17
Central Electricity Generating Board 50, 73
Centuary 65
Chadwick, M.J. 43

188 WASTE AND POLLUTION

chalk excavations 18
Chernobyl 151
Chew Valley 99
China 12
china clay 52
chloroflurocarbons (CFCs) 133, 143
Chough 25
Clean Air Act 1956 71
Cliff Rigg, Grear Ayrton 22
Coal fired power stations 79
coal mining 44
Cockroach, Common 32
Cocksfoot 39, 65
co-disposal 61
colliery spoil
Colorado beetle 117
Commons Select Committee on the Environment 55
Concorde 143
coniferous forest 104
Control of Pollution Act 1974 (COPA) 56
Control of Pollution (Special Wastes) Regulations 1980 56
Coot 49
copper arsenite 115
Corn Cockle 112
Cornwall 52
Cotswolds 50
Couch Grass 78
Coulson, J.C. 27
Country File (BBC) 22
County Councils 15
County Durham 44
Cowthick, Northants 22
Coxhoe tip, Co. Durham 28
Cranbrook, Lord 41
Creeping Bent 38, 53, 65
Creeping Buttercup 38
Creeping Thistle 39, 48
Cricket (insect) 32
crow 11
Curled Dock 39
Curlew 103
cyanide 62

Daphnia 98
Darlington, Arnold 13, 35

Daubenton's Bat 42
Davis, B.N.K. 54
DDE 119
DDT 111
decay 9
deoxygenation 88, 123
Deposit of Poisonous Waste Act (1972) 55
Deuteronomy 10
Devils bit Scabious 65
Devon 52
Dickens, Charles 70
domestic waste 15
Drakelow, river Trent 123
dung 9
dung beetles 9
Dunlin 95, 103
Dutch polders 42

Early Purple Orchid 65
Ecology of Refuse Tips 13
Edinburgh 12
Ehrlich, Paul 8
electrons 158
endosulphan 96
eutrophication 88
excrement 10
explosive mixtures 32

Farman, J.C., 145
Feral Pigeons 25
Field Maple 37
Field Vole 40
filter cake 36
firedamp 34
Flat Holm 28
fly tipping 16
Food chain 41
Forestry Commission 37
fossil fuel 10
foxes 17, 41, 118
Fragrant Orchid 66
Fremlin, J.H. 150
Freshwater Shrimp 91
Friends of the Earth 81
Frit Fly 113
fungicides 109, 114

Galloway, Scotland 104
gamma rays 150
Gemmell, R.P. 64
German Cockroach 32
Gilbert, O.L. 75
glass waste (cullet) 15
Glossop 77
glyphosate (Roundup) 113
Goat Willow 8, 45, 63
Gog Magog Hills, Cambridge 81
Golden Eagle 118
Golden Plover 103
Gorse 53
grazing marsh 21
Great Blackbacked Gull 25
Great Crested Grebe 49
Great Ouse 50
Great Willow Herb 49
Green, G.H. 26
greenhouse effect 129
greenhouse gases 35, 129
Greenpeace 60
Greenshank 103
Greig, J. 12
Guelder Rose 37
Gull, Common 25
gulls 24
Gunn's Bandicoot 16

Halley Bay, Antarctica 145
Hamilton, Victoria, Australia 16
Hare 40
Hares Down and Knowlestone Moor, SSSI 16
Harington, Sir John 12
Harwell Laboratory 56
Hawkes, Jacquetta 10
Hawkweed, Common 65
Hawksworth, D.L. and Rose F. 76
Hawthorn 37, 65
Hazardous Wastes 55
Heat, waste 14, 121
Hedgehog 41
Henton, J. 30
herbicides 109, 112
Herring Gull 25
Hertfordshire market gardens 12
high chimneys 77

Hiroshima 158
Holm Oak 37
hormesis 152
hormone weed killers 113
Horton, N., Brough T. and Rochard J.B.A. 26
House Sparrow 25
Huntspill, Somerset 21
Hynes, H.B. 90

insecticides 109, 115
ionising radiations 148
iron ore 47
Israelites 10

Jackdaw 25
Jack Snipe 103

Kaolin 52
Karin B 55, 61
Kestrel 25, 41
Knapweed, Common 65
Knotgrass, Common 35

landfill sites 17
Langford, T.E. 126
Lapps 158
Leach, G. 135
Lead 85
LeBlanc process 58, 64
lechate 32, 63
Leicestershire 41
Leisler's Bat 41
Lichens 74
Llanwllch Mere 22
Lobster 156
Loch Fleet, Galloway 106
London, 11, 122
London Brick Company 51
London particular 70
Lorenz, E. 154
Los Angeles, California 83
Love Canal 58
Lower Swansea Valley 55
Lyme Grass 39

magnesium deficiency 81
Malkins Bank, Cheshire 56

190 WASTE AND POLLUTION

Mallard 41, 95
Marsh Helleborine 65
Marsh Orchids 64
Mat Grass 66
Mauna Loa Observatory 131
Mayfly 91
MCPA 113
Meadow Grass 38
melanism 72
Mersey 96
metal waste 15
Metamorphosis of Ajax 13
methane 34, 58, 133
mica waste 53
mice 11
mining wastes 43
Ministry of Agriculture, Fisheries and Food 18
Mole 41
Moorhen 49
Mouse-ear Hawkweed 65
Moss, B. 99
Mudge, G.P. and Ferns, P.N. 27
Mugwort 39
Müller, P. 117

Naples 117
Nature Conservancy Council 5
nest boxes 42
Newcastle upon Tyne 75
NIMBY ("Not In My Back Yard") 17
nitrates 97
nitrogen 69
nitrogen oxides 83, 133
Noctule Bat 41
nomads 11
Norfolk Broads 99
North Yorkshire Moors National Park 22
Northamptonshire 47
Norway Spruce 80
Norwegian rivers 106
Nottinghamshire 47

Oak 48
Oat Grass 38
open-cast mining 39

Orache 38, 67
orchid hybrids 65
Orchids 18
organochlorines 116
organomercurial salts 115
organophosphorus compounds 116
Osprey 105
Ox-eye Daisy 39
Oxford clay 50
oxygen 69, 87
ozone 84
ozone layer 141

paper and cardboard 15
parathion 116
Paris Green 115
peat bog
Pentreath, R.J. 154
Peppered Moth 72
Perch 49
Peregrine 118
pesticides 109
phenolic waste 62
phenyl mercury 115
phenyloxyacetic acid 113
phosphate 97
phosphate stripping 100
pica (in children) 85
Pied Wagtail 25
Pintail 95
Pipistrelle Bat 41
Pitsea, Essex 19
Pitsea Hall Fleet 20, 41
plastic 15
Pochard 91
polychlor biphenyls (PCBs) 60
Poplar 37
porcelain industry 52
potato blight 114
pozzalans 63
Preseli, N.Wales 31
Prymnesium 99
pulverised fly ash (PFA) 50
Pyramidal Orchid 65
pyrite 44, 67

quarries 54
Queen Elizabeth I 12

INDEX 191

Rabbit 40
radioactive waste 14, 148
radon 69, 153
Ramsar Convention 21
Ramsey Heights, Huntingdonshire 50
rats 11, 40
Rat-tailed Maggot 91
Raven 25
recycle 9, 163
Red Clover 39, 53
Red Fescue 53
Red Kite 11
Red Spider Mite 119
Redshank 95
Reed 49
Reed Bunting 103
Reed Mace 49
refuse 9
Restoration of Land 43
Rhine 96
Ribbed Melliot 39
Ribwort Plantain 39
Roach 49
Rook 25
Rosebay Willowherb 48
Rossi, Sir Hugh 55, 59
Rough Meadow Grass 38
Royal Commission on Environmental Pollution 13, 35, 55, 61
Royal Society of Chemistry 111
rubbish 9
Ruff 103
Rye Grass 38, 53
Rye Meads Sewage Purification Works 103

Sallow 49
Salmonella 28, 98
Salt Marsh Grass 67
sand and gravel 49
Sandpiper 103
sanitary landfill site 6, 23
Savannah River, S. Carolina 125
scabies 118
Sea Aster 67
seaweed 156
Sedge Warbler 103

Sellafield 156
Serotine Bat 41
Sessile Oak 67
Sewage farms 13, 102
sewage sludge 81
sewage treatment 13, 89
sewage works 25
Seward, M.R.D. 76
Shaw, P.J. 37
Sheep's Fescue 65
Sheep's Sorrel 66
Shelduck 95
Shipworm 127
Shrew, Common 40
silage effluent 95
Silver Birch 48, 65
simazine 111
slate industry 54
Sludge Worm 91
Small-leaved Lime 37
Smith, R.A., First Alkali Inspector 78
smokeless zones 15, 71
smog deaths, 1952 70
Solvay or Ammonia Soda Process 65
"Somernet" mobile cover net 30
Somerset County Council 21
South Wales 44
Sparrowhawk 118
Spear Thistle 39
Spotted Orchid 66
Spurrey 67
Steep Holm 28
Stone age man 10
Stone Fly 91
strip mining 45
sulphur dioxide 72
Sunday Times newspaper 91
Supersonic aeroplanes 143
Sweden 156
Sycamore 37, 58

Tamarisk 37
tar spot disease 74
Teal 95
Teasel 39
Temminck's Stint 103
tidal power 136

Timothy 38
Tollund man 11
transfer station 22
trash 9
Tree Lupin 53
troposphere 69
Tufted Duck 95
typhus 117

ultra-violet radiation 141
urine 11

Vale of Belvoir 44
Valroff, J. 8
vegetable and putrescible matter 15
Verseilles 12

waldsterben 79
Warwickshire 26
Wash, The (Lincolnshire) 19
washing soda 64
waste 9
waste heat 14, 121
water closets 13
Water Mint 49

water pollution 57
Watt Committee on Energy 135
Wavy Hair Grass 67
Weasel 40
Westminster Abbey 82
Wheat bulb fly 115
Whimbrel 103
White Clover 39, 53
Wild Carrot 39
Wildlife and Countryside Act 1981 8
Wild Oat, Common 38
Willow 37
Wilson, D.G., Smith, E.F. and Pearse, K.W. 56
wind power 136
Wood Mouse 40
Worcestershire 26

X-rays 150

Yarrow 39
Yellow Iris 49
Yorkshire 44
Yorkshire Fog 39